Evolution and Modification of Behavior

Evolution and Modification of Behavior

Konrad Lorenz

The University of Chicago Press

Chicago / London

THE UNIVERSITY OF CHICAGO PRESS, CHICAGO 60637
The University of Chicago Press, Ltd., London

Published 1965. First Phoenix Edition 1967. Printed in the United States of America

80 79 78 77 76   11 10 9 8 7

Contents

1

Introduction

What is preformed in the genome and inherited by the individual is not any "character," such as we can see and describe in a living organism, but a limited range of possible forms in which an identical genetic blueprint can find its expression in phenogeny. The way in which this takes place has been well enough analyzed by the sciences of genetics and phenogenetics, at least in some paradigmatic cases. The term "innate" should never, on principle, be applied to organs or behavior patterns, even if their modifiability should be negligible. N. Tinbergen and many other ethologists writing in English, have, therefore, altogether ceased to use the word innate, or, to be more exact, they continue to apply it only to differences appearing between behavioral or morphological characters developed by individuals reared under identical environmental conditions.

There are two debatable points about this way of applying the term: the first question is how we define "character," and the second is whether the conception of a "strictly identical environment" is operationally valid. Nobody, however, doubts that it is the most convincing proof of heredity when hybrids of known parentage reared under identical conditions develop characters whose differences from each other are clearly correlated with those existing between the parent species and concur with the known laws of inheritance, as for instance, in the dabbling duck hybrids bred by von de Wall (1963). In the F_1 generation, these birds developed patterns of courtship behavior that were either roughly intermediate between those of the parental species, or showed ancestral characters existing in a large number of other ducks but in neither of the parental species. In the F_2 generation, the motor patterns of the individuals differed widely, showing many new combinations of recognizable characters found in the grandparent species.

The genetic hybridization experiment is not, however, the only occasion on which we have need of a term, and, commendable though semantic purism may be, it leaves us without a word denoting an indispensable concept. We did mean something when we spoke of an innate motor pattern or an innate releasing mechanism, and it is not just semantic carelessness if some of us tend to use the rather cumbersome expression, "what we formerly called innate." What indeed? The obvious need for a term is a sure indication that a concept which corresponds to something very real does exist. The aim of this book is to find out what this reality is and to define this concept. This attempt will imply first, a detailed critique of some widely held views concerning the concept of the "innate" and second, a discussion of the value and the limitations of the deprivation experiment, of the assertions which it enables us to make, and of the methodological rules that must be observed in its application.

2
Theoretical Attitudes toward the Concept of the "Innate"

Three distinct attitudes regarding the concept of "what we formerly called innate" are sufficiently important to merit discussion. I believe that I can show some serious logical and biological fallacies in the first, which is held by the majority of American psychologists. The second is the one taken by most modern English-speaking ethologists who, in my opinion, have lost hold of an all-important concept, partly from overcaution, partly because they wished to compromise with behavioristic critique, but mostly in consequence of a rebound phenomenon on discovering some errors in the "naïve" attitude which Heinroth and other older ethologists assumed toward the concept of the innate and which must also be criticized in this book.

Most American psychologists, whom for brevity's sake and with some incorrectness, I shall here subsume under the concept of "Behaviorists," maintain, more or less unanimously, that the "dichotomy" of behavior into innate and learned elements is not "analytically valid." This statement is supported by two arguments. The first of these asserts that the dichotomy is only the result of "begging the question" as hitherto the only definition of the "innate" is that which is not learned and vice versa. Hebb (1953) writes, "There must be great doubt about the unity of factors that are identified only by exclusion" and "I urge that there are not two kinds of control of behaviour, and that the term 'instinct,' implying a mechanism or neural process independent of environmental factors, and distinct from the neural processes into which learning enters, is a completely misleading term."

The second argument, emphasized by Lehrman (1953), contends that, even if the existence of behavior elements which are independent of learning cannot be entirely ruled out, the concept of innate behavior is nevertheless without heuristic value because it will never be practically possible to exclude the participation of learning in the early ontogenetic processes in the egg or *in utero,* which are inaccessible to observation.

Clearly distinct from the behaviorist's attitude, yet superficially somewhat similar, is that taken by Tinbergen (1955, 1963) and a great number of other modern ethologists. Although they have dropped the word "innate" for the terminological reasons already mentioned, they are, of course, fully aware that there really are two entirely independent mechanisms affecting adaptation of behavior: the process of phylogeny which evolves behavior as well as any other structural and functional organization, and the processes of adaptive modification of behavior during the individual's life. In spite of agreeing, on principle, with this "dichotomy," these scientists take the attitude that practically all behavior, down to its smallest units, owes its adaptedness to both of the above adaptive processes and that the types of behavior which we formerly described as "innate" and "learned" represent only the two extremes in a continuum of gradation in which all possible mixtures and blendings of the two sources of adaptation can be found. That these two extremes actually occur and do so with a surprising frequency is fully recognized but explained by the tentative and rather arbitrary assumption that the extremes, being of particular survival value, are favored by selection pressure more than the intermediate forms.

As a consequence of this attitude, any attempt to separate phylogenetically and individually adapted characters and properties of behavior, either conceptually or in the course of practical experiments, must necessarily be considered as hopeless and devoid of sense, as any trait of behavior, however minute, is automatically regarded, on principle, as being influenced by both factors achieving adaptation. This highly dangerous opinion is often supported by an argument based on some practical limitations inherent in the deprivation experiment (p. 83).

The third and most controversial attitude rests on the assumption that instinctive and learned behavior "come in chunks" which can be clearly separated from each other, as implied by *Instinkt-Dressur-Verschränkung* (intercalation of fixed pattern and learning) which I proposed more than thirty years ago. On the basis of this assumption, the obvious trend in the evolution of behavior in the direction of greater plasticity and increasing influence of learning and insight has to be regarded at least as much as a consequence of reduction and disintegration of innate fixed patterns as of higher development of those functions which, in the individual's life, affect adaptive modification of behavior.

3 Critique of the First Behavioristic Argument

It is simply not true that "what we formerly called innate" and "what we formerly called learned" can only be defined one by the exclusion of the other. Like another fundamental error, which will be criticized in the discussion of the second behavioristic argument, this fallacy arises if one forgets that it must never be regarded as a product of chance or taken as a matter of course when behavior is found to be adapted to a corresponding point of the species' environment. "Adaptation" is the process which molds the organism so that it fits its environment in a way achieving survival. Adaptedness is always the irrefutable proof that this process has taken place. Any molding of the organism to its environment is a process so closely akin to that of forming, within organic structure, an image of the environment that it is completely correct to speak of information concerning environment being acquired by the organism.

There are only two ways this can happen. The first is the

interaction between the organism and its environment. In this process it is the species which, by means of mutation and selection, achieves adaptedness insuring survival. All the complicated structures and functions of the chromosomes, including mutation and sexual reproduction, are a mechanism evolved in the service of the function of acquiring and storing information on the environment. A very close functional analogy exists between trial-and-error learning and the way in which part of the progeny is risked in the "experiment" of mutation, a part which is cautiously measured by a mutation rate which does not jeopardize the survival of the species and, therewith, the hoard of information already stored. Probably one of the chief functions of sexual reproduction is to disseminate new information quickly among a population.

Campbell has called attention to the fact that the procedure by which a species attains information about its environment is virtually identical with that of pure induction and is devoid of the deductive processes which guide human experimentation. The information thus gained is stored in the genes, which for this reason have been aptly described as "coded information" by geneticists.

The organization which achieves all this obviously must have been evolved on our planet at a very early stage in the history of life. We know that all higher animals and plants are descended from organisms which had already "invented" a nucleus with chromosomes performing these functions. With the advancement of biochemical knowledge concerning analogous processes in the very lowest organisms, in bacteria and viruses, we are seriously confronted, however, with the question whether the origin of these mechanisms for acquiring information might not be identical with the origin of life itself.

The second way by which information on the environment can be fed into the organic system is the interaction between the individual and its surroundings. Every response-eliciting stimulus impinging on the organism represents information on environment being received by the animal, even if this information only determines when and where a certain mechanism of behavior is activated without, however, affecting any adaptive change in the mechanism itself. All "reflexes," releasing processes, etc. as well as all orientation responses, or taxes, belong to this category.

Among all the mechanisms serving the gain of immediately exploited information, those achieving orientation in space stand apart insofar as they determine not only at what moment but also in what direction a motor pattern is to be discharged. These mechanisms imparting instantaneous temporal and spatial information are, in their more highly developed forms, largely identical with what is called intelligence in common parlance. Subjectively, their successful function is accompanied by the experience of "insight." They very often produce "learning" and frequently are its prerequisite. They are, however, totally independent of learning; they function even in the lowest unicellular organisms which do not learn.

Particularly on the basis of one constitutive property, these mechanisms must be conceptually distinguished from learning. This has been pointed out by Russell (1958–61). The processes that are under discussion here achieve an instantaneous adaptation of behavior to the environmental requirements of the moment; they do not cause a modification in the mechanism of the response but are themselves the function of the highly differentiated, phylogenetically adapted mechanisms, which, incidentally, are largely identical with what Pavlov called unconditioned reflexes.

Learning, on the other hand, is closely akin to the phylogenetical processes already discussed in its all-important function of not only acquiring, but also of storing, information. We do not know as yet how this storing is achieved. It might be done by adaptively modifying neural structure, or, as some biochemists tentatively like to assume, by coding information in chain molecules in the way phylogenetical information is retained. But whatever the mechanism, adaptive modification of a function cannot take place without corresponding modification of the structure underlying it. The enormous difference in their respective time scales and in their physiological mechanisms notwithstanding, the functional analogies between the phylogenetical and the individual processes of acquiring and storing information are such that Russell seems well justified in creating a concept encompassing both.

Our absolute ignorance of the physiological mechanisms underlying learning makes our knowledge of the causation of phyletic adaptation seem quite considerable by comparison. Our conviction that Darwin's concepts correspond to realities is based on facts that grow more solid year by year and have lately received an unshakeable foundation from the modern results of biochemistry. Hence, when we talk of phylogenetic adaptation and the processes underlying it, we use concepts based on and defined by causal insights. When we put phylogenetic adaptation of behavior in antithesis to learning, we must bear in mind that the two concepts are arrived at in different ways. Learning, unlike phylogenetic adaptation, can be defined only descriptively and functionally and, moreover, only by what Hassenstein (1954) has termed an "injunctive" definition, that is, by an enumeration of a number of part-constitutive properties. Most biological concepts, including that of life itself, can only be defined injunctively. Every single one of its part-constitutive properties: assimilation, metabolism, growth, and reproduction can be found in some non-organic model; all of them together make life. Hence, every injunctive concept has a hard core in which its applicability is unquestionable, although, on its margins, it merges into neighboring concepts without a strictly delineable border. A purely injunctive and, since we know nothing about its physiological causation, purely functional definition of learning must lack any sharp delineation. The core of the concept will be represented by classic conditioning and other "higher" forms of learning, and marginally it will merge imperceptively into more lowly forms of adaptive modification of behavior.

On the receptor side, more highly specific habituation (see also p. 48) via sensory adaptation will form a gradation leading to processes that one would not call learning at all. On the motor side, acquiring skills through practice will grade into more simple forms of facilitation by exercise and further into the unspecific effect of activity which, by preventing the atrophy otherwise seizing many inactive structures, forms a simple prerequisite for the normal phenogeny of the organ. Although well aware of all this, I propose, for the purposes of this book, to define learning in a much too comprehensive way, including all adaptive modifications of behavior. Any workable definition of learning must unconditionally contain, as its most indispensable part-constitutive property, the character of survival value or adaptiveness, as Thorpe pointed out in 1956.

Whatever else learning may be, it certainly is an adaptive modification of behavior, and its adaptiveness, that is, its ability to adapt behavior, needs a causal explanation. There is an infinitesimally small chance that modification, as such, is adaptive to the particular environmental influence that happened to bring it about. Indeed, this chance is not greater than that of a mutation being adaptive. Geneticists rate this likelihood at about 10^{-8}. Whenever we do find a clearly adaptive range of modifiability, for instance, when we find that the thickness of fur is adaptively modified by influences of the climate in very many mammals or that plants stretch farther upward the less light they get, we know that these achievements of adaptation are not exclusively due to environmental influences, but just as much to a very specialized range of modifiability which has been selected for in the pre-history of the species.

The more complicated an adapted process, the less chance there is that a random change will improve its adaptedness.* There are no life processes more complicated than those which take place in the central nervous system and control behavior. Random change must, with an overpowering probability, result in their disintegration. Blithely assuming that "learning" (whatever that may be) automatically achieves adaptive improvement of behavior mechanisms implies neither more nor less than the belief in a prestabilized harmony between organisms and environment. The amazing and never-to-be-forgotten fact is that learning does, in the majority of cases, increase the survival value of the behavior mechanisms which it modifies. The rare instances in which this survival function miscarries serve to illustrate rather than to negate this—but this fact itself demands an explanation.

*Supposing that an amateur mechanic would change, randomly or on the basis of insufficient information, anything in the mechanism of the engine of his "thoroughbred" sports car, the chances of his improving its function are negligible.

I regard it as highly paradoxical that many American psychologists not only seem to forget this, but actually stigmatize as "preformationistic" the theories of ethologists who take into consideration that some information underlying an individual's behavior has indeed been "preformed" by the species. This type of psychologist probably finds it so very easy to forget the questions of survival function and phylogenetic origin of behavior in general and of learning in particular, because their experimental setups deviate from the natural surroundings of the species investigated to such an extent that it becomes all too easy to overlook the survival function of behavior altogether and, therewith, the selection pressure which caused its mechanisms to evolve. To anyone tolerably versed in biological thought, it is a matter of course that learning, like any function of comparably high differentiation and survival value, must necessarily be performed by a very special mechanism built into the organic system in the course of its evolution. The indubitable fact that every learning mechanism is phylogenetically evolved is in no way contradicted by the other fact that a learning mechanism is evolved to exploit individual experience. Both facts together inescapably raise the question of how this learning mechanism achieves the task of choosing among innumerable possibilities of behavior those for reinforcement which develop positive survival value and, for extinction, those which are detrimental to the individual or the species.

In the days of classic behaviorism, it was assumed that the fulfillment of primary bodily needs tended to reinforce, and that any bodily damage tended to extinguish the type of behavior which happened to precede these effects, thus directing the learning process toward survival value (Thorndike, 1911). Although still assuming that the organism, in some unexplained manner, should "know" what is good and bad for it, this theory not only bears in mind that an explanation is needed for the survival value of learning, but even tentatively offers one which is not too far from reality. There must be some neural mechanisms which, on their afferent side, are rooted in the homeostatic cycles of elementary vegetative and metabolic processes, and which report to the central nervous system any deviation from that particular steady state which is desirable for survival. In our own subjective experience, the report of such a disturbance is correlated with the diffuse feeling of illness which cannot be localized. With the mechanism of pain, obviously less elaborately organized on its receptor side, the accent lies on reporting the localization of the disturbance. The survival function of both organizations indubitably lies in the strong extinguishing and reinforcing effects produced by the waxing and waning of their reports.

The function of a very generalized "illness receptor" could perhaps furnish a tentative explanation for the amazing results which Richter (1954) obtained when he supplied his experimental animals with foodstuffs disintegrated into their component parts and left it to the subjects to resynthesize a balanced diet. Even when the amino acids of the necessary proteins were offered separately, the rats, as subsequent weighings showed, took just the right percentage of each constituent. Animals deprived of the adrenal cortex proved to be able to compensate for the disturbance which the operation caused in their salt metabolism by eating a correspondingly increased amount of NaCl.

The most intriguing question posed by these findings is: from whence does the organism obtain the information telling it what chemicals are needed in its metabolism at a given moment, and how are these to be recognized when encountered in the environment. We know of one case at least in which a special releasing mechanism achieves the recognition of a needed chemical: birds lacking calcium will peck at and eat any white, hard, and crumbling substance, regardless of its chemical composition, obviously guided by visual and tactile stimuli rather than chemical ones. I have known birds to poison themselves by eating carbide. It is, however, enormously improbable that similar phylogenetically evolved perceptual organizations are lying in readiness for each of the very special needs selectively supplied in Richter's experiments. His rats were, in all probability, the first ones in the evolutional history of their species to synthesize protein from its component amino acids. It is a tempting, if untested, hypothesis that the animal tentatively eats just a little of each substance offered to it and forms an engram of "how it feels afterward." This provisional assumption is supported by the fact that omnivorous animals invariably at the first encounter eat very little of any unknown food. Similar considerations might apply to the compensation of salt intake by adrenalectomized rats.

Hull's (1943) important and indubitably correct assertion that all relief of tension acts as a reinforcement on the behavior which preceded it also presupposes the function of a built-in mechanism which, with a similarly wide range of applicability as the one just discussed, is able to direct learning toward survival value in the most varied environmental situations. Appetites and aversions, regarded by Craig (1918) as diametrically opposed types of behavior, have one structural property in common. In both cases, "purposive" behavior modifiable by learning is directed at a reinforcing external and internal stimulus situation which results in a marked relief of tension. The physiological difference lies only in the provenience of this tension. In the case of appetitive behavior, it is caused by the internal generation of endogenous stimuli pertaining to the instinctive movements which appetitive behavior is striving to "discharge." In the case of aversion, the tension is directly caused by a disturbing external situation. The organism tries to rid itself of this situation by "appetitive behavior directed at states of quiescence," as Meyer-Holzapfel (1940) more objectively describes aversion.

All the teaching mechanisms hitherto mentioned contain phylogenetically acquired information that tells the organism which of the consequences of its behavior must be repeatedly attained and which ought to be avoided in the interest of survival. This information is preponderantly localized in the perceptual organizations which respond selectively to certain external and/or internal configurations of stimuli and report them, with a plus or minus sign added, to the central mechanisms of learning.

As in the teaching mechanism just discussed, this information is frequently couched in very generalized, "abstract" terms. The organization of perceptual patterns is always confronted with the alternative of achieving either high selectivity or general applicability, or of compromising between these two properties which, in their higher degrees of differentiation, obviously cannot be achieved simultaneously. In perceptual patterns in which, as in the reinforcing mechanisms just discussed, a very general applicability is imperative, it can only be attained at the cost of low selectivity. In other words, the organism is unavoidably exposed to the dangers incurred by a correspondingly great probability of its perceptual reinforcing mechanisms miscarrying. In many omnivorous animals, for example, a mechanism exists that causes them to prefer food with a minimum content of fiber and a maximum of sugar, fat, and starch. In the "normal" conditions of wild life, this phylogenetically adapted releasing mechanism is of obvious survival value, but in civilized man it gives rise to a search for supernormal objects, the addiction to which actually amounts to a vice detrimental to health (e.g., white bread, chocolate, etc., which cause constipation and obesity in millions). In closely analogous manner, another important teaching mechanism can miscarry, that is, a mechanism which acts on the generally reliable "hypothesis" that a situation affording relief of tension is also one desirable in the interest of survival. As is to be expected, the mechanism also responds to all drugs causing relief of tension and thus conditions men and cats (Masserman, 1943) to become alcoholics and tranquilizer addicts.

The limited probability of a behavior mechanism thus miscarrying is no threat to the survival of the species, and it is still less of an argument against the well-based assumption that the mechanism in question is, under normal circumstances, of great survival value. Quite the contrary; the disturbance offers valuable insight into the causal structure by whose function that survival value is achieved.

It is childishly easy to construct thought models or machines which will "learn something" that is to say their functional properties will get altered by previous functioning. It is even difficult to devise any complicated electronic apparatus which is not susceptible to such changes, but it is immensely difficult to invent even a thought model of a system in which the change of function caused by functioning will always be in the direction of supporting, regulating, and repairing that system. The central problem of all reinforcing and/or extinguishing mechanisms lies in their content of innate information telling them what is "good" and what is "bad" for the organism. It is only on the assumption of a prestabilized harmony between the organism and its environment that one can dispense with an explanation for the survival value of learning and with an investigation of the innate information contained in the teaching mechanisms achieving it.

The "dichotomy" of behavior into "innate" and "learned" components is indeed misleading but in a sense directly opposed to that implied by Hebb. The assumption that learning "enters into" every phylogenetically adapted behavior mechanism is neither a logical necessity nor in any way supported by observational and experimental fact. Although practically every functional unit of behavior contains individually acquired information in the form of a stimulus-input which releases and directs reflexes, taxes, etc. (p. 9) and thus determines the time and the place at and in which the behavior is performed, it is by no means a logical necessity that adaptive modification of behavioral mechanisms is invariably concerned in the survival function of these short-notice responses. These responses, in their very essence, are the opposite of "what we formerly called learned," being, indeed, as already mentioned, what Pavlov has termed "unconditioned reflexes."

On the other hand, it is, for the reasons already expounded, an inescapable logical necessity to assume that learning, like any other organic function regularly achieving survival value, is performed by organic structures evolved in the course of phylogeny under the selection pressure of just that survival value. All observational and experimental evidence goes to confirm this assumption. None contradicts it.

Besides the two channels represented (*a*) by the adaptive processes of evolution, and (*b*) by individual acquisition of information, no third possibility exists for information being "fed into" the organic system. Whenever and wherever a behavior pattern shows structure adapted to a corresponding point of environment, one or both of these processes must be assumed to have taken place; and it is, at least in principle, always possible to ascertain what part each of them has played. No biologist versed in phylogenetic and genetic thought would ever dream of dropping the two concepts of phyletic adaptation and of adaptive modification just because, in many cases, both are concerned in molding organic structure and/or function to fit environment. The fact that, in the case of the phenocopy, either of the two processes can achieve a result indistinguishable from that of the other, has never been considered a reason for dropping the conceptions of the two processes as being "not analytically valid."

Nor must the functional analogies which exist between the two types of acquiring information mislead anyone into thinking them to be physiologically the same. Indeed, we know that they are not. Not even the principle of "trial and error" is quite the same in both cases. It is only individual learning that is able to gain information from its errors. The hit-and-miss method of mutation and selection gains only by its successes and not by its failures and continues blindly to produce those mutants that have proved unsuccessful millions of years ago. Still, there are sufficient reasons for uniting both under one functionally defined concept as Russell (1958–61) has done.

Physiologically, however, it is an entirely different process when, on the one hand, a species "experiments" by mutation or Mendelization and "records" results by one animal surviving and the other dying, and when, on the other hand, an animal "experiments" by doing this and that successively and "records" by forming a neural engram of what has brought about a reinforcing stimulus situation and what has not. For one thing, the amount of time needed for each of these processes are different by quite a few powers of ten. In addition, the coded information stored in the genes, and also its decoding in ontogenetic development, is different from the information stored in the nervous system and its decoding by whatever happens in the releasing of a conditioned response. No similarity of results and no difficulty in the practical task of analysis can ever give us exemption from the scientific duty to trace back, to one or to the other of the two sources of information, any single point in which behavior can be shown to be adaptively molded to a corresponding point in the environment of the animal.

Hebb, in his criticism of ethology, says that it would be hard to exaggerate the importance attributed by ethologists to the distinction between the innate and the learned, implying, of course, that we overestimate this importance. I could not agree more with the statement and less with the implication. If, as physiologists of behavior, we intend to analyze the causality of adapted behavior at all, I do not see how we can pursue this object without ascertaining the source of information underlying adaptation. There are also practical reasons which make the distinction important. For instance, the belief that human aggression is based not on phylogenetic adaptation but on learning implies a tremendous underassessment of its dangers. Hitherto this belief has only led to the production of thousands of intolerably aggressive non-frustated children, but it may lead to much worse things.

Lastly, a very deep misunderstanding of biological ways

of thinking is contained in the sentence by Hebb (p. 14), expressing the opinion that "instinct [in the sense of a neural organization according to Tinbergen, 1952] implies a nervous process or mechanism . . . which . . . is different from those processes into which learning enters." We certainly do not believe that there is just one process into which learning does not enter. Even if we adhere to the superlatively comprehensive definition of learning which I proposed on p.11 and which includes all adaptive modification, we still know dozens of very complicated behavior mechanisms whose adaptedness is entirely based on phylogenetically acquired information. The computing mechanism which enables a starling to deduce the points of the compass from the motion of the sun across the sky (a sun which the bird has never seen; Hoffmann, 1952), the complex feedback mechanism which enables a mantid to aim its grabbing movement unerringly at the prey (Mittelstaedt, 1957), the unvarying inherited courtship movement of a drake which releases a specific answer in a duck of the same species (Heinroth, 1910), the "internal clock" which prescribes rhythmical recurrence of activities in so many animals (Aschoff, 1962), all optomotor responses, and so on are all "nervous processes or mechanisms which are different from those into which learning enters." They are as different from each other as a tooth is different from a bone or a kidney and exactly for the same reasons. We do not know how many more such mechanisms or processes do exist—all different from and independent of each other.

4 Critique of the Second Behavioristic Argument

This argument vastly overrates the amount of information that can be gained by the embryo during intrauterine life or in the egg. This error indubitably arises from the same fundamental fallacy which underlies the argument already discussed, namely, from forgetting that adaptedness to environment can never be a coincidence but must necessarily have a history explaining it. If Lehrman (1953) gives serious consideration to the assumption that a chick could learn, within the egg, considerable portions of the pecking behavior by having its head moved rhythmically up and down through the beating of its own heart, he totally fails to explain why the motor pattern thus individually acquired should fit the requirements of eating in an environmental situation which demands adaptedness to innumerable single givens as exactly as it does. It also remains unexplained why only certain birds peck after hatching, while others gape like passerines, dabble like ducks, or shove their bills into the corner of the mouth of the parents as pigeons do, although they all, when embryos, had their heads moved up and down by the heartbeat in exactly the same fashion. The more than wonderful adaptedness is passed over as a matter of course, in spite of the fact that it would require truly astronomical numbers to express the improbability of its arising by chance.

Adaptedness to environment can only be regarded as a matter of course by a man who, like von Uexküll, assumes a prestabilized harmony between organism and environment. Barring this assumption, anybody thinking it possible that the bird learns, within the egg, behavior which fits exigencies not encountered until later in life, automatically has to assume the existence of a teaching apparatus which contains the phylogenetically acquired information concerning those exigencies. Neither Kuo (1932) nor Lehrman (1953) made this assumption.

Thus, paradoxically, the notion that the organism could, within the egg or *in utero,* learn behavior specifically adapted to later environment implies preformationism. An amusing double paradox lies in the fact that this preformationism is the penalty which some American psychologists incurred by trying to avoid, at all costs, the concepts of survival value and phylogenetic adaptation for no other reason than that they regarded them as "finalistic." Of course, they are not finalistic in the least. If a biologist says that the cat has crooked, pointed claws "with which to catch mice," he is not professing a belief in a mystical teleology, but succinctly stating that catching mice is the function whose selection pressure caused the evolution of that particular form of claws.

There is still a considerable element of truth contained in the second behavioristic argument. There are learning processes which can possibly take place within the egg or *in utero.* Prechtl (1958) has demonstrated some in the human fetus. The deprivation experiment, which will be discussed later, permits assertions concerning the innate information about only those environmental givens with which the subject was specifically prevented from having experience. So it is practically never possible to state that all the information underlying the adaptedness of a whole functional unit is phylogenetically acquired. Still, in very many cases the fault which would be contained in such a not-quite-justifiable assertion would be very much smaller than is often assumed on the basis of the above-mentioned unintentional preformationism.

In very many cases, the adaptedness of behavior can be traced back to innate information, even without performing any deprivation experiment. A young male salticid spider which, after molting for the last time, approaches a female must neither mistake another species for his own, nor must he perform the signaling of his specific courtship dance in any other way than the one to which his female responds; otherwise, he would be eaten by her immediately. He has no opportunity in his short life to gain any information about what a female of his own species looks like, nor what movements he must perform to inhibit her feeding reactions and to stimulate her specific mating responses.

A young swift reared in a narrow cave in which it cannot extend its wings (far less beat them up and down), in which it cannot attain a sharp retinal image as the farthest point of the cave is much nearer than its shortest focusing distance, and in which it cannot gain any experience on parallactic shifting of retinal images, nevertheless proves to be perfectly able on the very moment it leaves the nest cavity to assess distances by the parallactic shift of the object's images. It can also cope, in its rapid flight, with all the intricacies of air resistance, upcurrents, turbulence, and air pockets and can "recognize" and catch prey, and finally effect a precise landing in a suitable place. The information implicitly contained in the adaptive molding of all these forms of behavior to the environmental givens to which they indubitably do fit would fill many volumes. The description of the innate distance computers alone would contain whole textbooks of stereometry and that of the responses and activities of flying, an equal number of data on aerodynamics.

If one assesses, even ever so roughly, the amount of information which quite indubitably is transmitted by way of the genome and compares it with what (even conceding the most unlikely possibilities) could have been acquired in individual life, the proportion is astounding. What the swift could have learned about stereometry is practically nil, even when we attribute quite superhuman learning powers to the bird. In its ability to fly, it could have learned even less than a human could learn about skiing on an indoor ski course. All the information it could have acquired individually could concern only its own body and a few basic and ubiquitous laws of physics. For instance, it could have learned to innervate synergistic and antagonistic muscles alternately. It could, from tactile experience, have deduced the first law of physics and know that two things cannot be in the same place simultaneously. Even overlooking the unlikelihood of simple motor coordinations being acquired—we know otherwise from the work of von Holst, Eibl-Eibesfeldt, and others—and the improbability of the bird's being able to apply what it has learned by tactile experience to its orientation while flying in the air and even if we attribute to the bird superhuman ability to learn, we know the proportion of learned information which could possibly enter into the responses and these activities is infinitesimally small. The naïve ethologist's assertion that they are "completely innate" is indeed less inexact than the statement that a steam locomotive or the Eiffel Tower are built entirely of metal. In other words, it attains an exactitude rarely reached by biological assertions.

5 Critique of the Modern Ethologists' Attitude

What I propose to discuss here is mainly the assumptions already mentioned in the introduction: that "what we formerly called innate" and what we formerly called "learned" represent only the extreme ends of a continuum of insensibly graduated mixtures between the two, and that all behavior, down to its smallest elements, owes its adaptedness to both processes. I think I can show that this assumption is not only bad strategy of research but completely unfounded and in all probability false.

In discussing the first behavioristic argument, I have attempted to demonstrate the fallacy of treating the "innate" and the "learned" as opposed, mutually exclusive concepts. I tried to show that, while all learning is performed by mechanisms which do contain phylogenetically acquired information, no reasons exist for assuming that individually acquired information "enters into" every kind of phylogenetically adapted behavior. For millions of years our planet has been inhabited by creatures possessing quite elaborate behavior patterns which owed none of their adaptedness to any of the higher and typical forms of learning. We are certain that habituation and its counterpart, sensitization, are the only forms of adaptive modification of behavior ever found in protozoa and in organisms with a diffuse nervous system. When, aeons later, at a comparatively high level of neural organization, modifiability of behavior began to increase, a tremendous selection pressure must have been brought to bear on the development of mechanisms adaptively modifying individual behavior. The intensity of this pressure can be inferred from the omnipresence of its effect in all phyla of organisms which, independently of each other, have evolved a centralized nervous system.

Everything that has been said in criticism of the first behavioristic argument is a reason for not assuming an unlimited, diffuse mutual permeability and mixability of phylogenetic adaptation and learning. Only confirmed "preformationists" can doubt the fact that any specialized adaptive modifiability of behavior, such as we find in vertebrates, arthropods, and cephalopods, can never be a product of chance, but has to be regarded as the function of a nervous apparatus which is highly organized and which owes its origin, as all organization does, to the process of "pure induction" (p. 8) conducted by the evolution of every species. To assume diffuse permeability of phylogenetic adaptation and learning implies the assumption of an infinite number of such organizations, which is strictly contradicted by the hackneyed fact that there are, for each animal species, only limited numbers of highly specific stimulus situations that will act as reinforcements.

The notion that learning or any other change of behavior achieving survival value could possibly be the function of a non-specifically organized and programmed aggregation of neural elements, is absolutely untenable.

It is a matter of surprise and concern to me that this fallacy has been voiced in "Behaviour" with great self-confidence and without arousing the least contradiction on the part of English-speaking ethologists. Ethology loses its character of a biological science if the fact is forgotten that adaptedness exists and needs an explanation. There is no hope of gaining greater "exactitude" by shedding biological knowledge and clinging to allegedly operational concepts. All conceptualizations employed in the scientific study of nature are fundamentally operational, and if the operations underlying the concepts used in phylogenetics are complicated and presuppose considerable knowledge, this furnishes no excuse for neglecting important and well-known facts. It is a great error to believe that "exactitude" can be gained by restricting research to one experimentally —and mentally—simple operation. The damage done by this type of procedure is best illustrated by Jensen's paper "Operationism and the Question 'Is This Behavior Learned or Innate?' " (1961).

Jensen argues in the following manner:

> The possibility of differences in behavior being neither innate (selected) nor learned (trained) but something else (produced by other operations) allows the separate classification of various types of operations or antecedent differences to which behavioral differences can be attributed.

(A list of such operations, including food schedule, hormones, and even nervous system lesions, etc. follows.)

> Since behavioral differences may be attributed to effects and interactions of many factors, the original question can be rephrased in an unrestricted way. Instead of: 'Is this behavior

innate or learned?' we can ask: 'What causes this difference in behavior?' So asked, the question becomes a matter for research instead of argument.

There is not and there cannot be any argument about the fact that the Kuenzers' (1962) young *Apistogramma* responded selectively to the configurational key stimulus of a yellow and black pattern by the specific activities of following the parent and that a mother *Apistogramma* is indeed striped black and yellow. Nor is it a matter of argument that a stickleback responds to the key stimulus "red below" by performing the motor patterns of rival fighting and that a male stickleback is indeed red on the ventral side. Only if we should forget these central facts of adaptedness, and only then, should we have to resort to the desperate strategy of research which Jensen recommends to us, unless we should prefer to give up as hopeless any further attempt to analyze behavior.

Non-adaptive differences in structure and behavior are of but secondary interest to the biologist, while they are the primary concern of the pathologist. A living organism is a very complicated and very finely balanced system, and there is precious little about it that cannot, at least on principle, be understood on the basis of phylogenetic adaptation and adaptive modification. Neither in an animal's body nor in its behavior are there many characters that may be changed by "other operations" without leading to destruction.

As students of behavior, we are not interested in ascertaining at random the innumerable factors that might lead to minute, just bearable differences of behavior bordering on the pathological. What we want to elucidate are the amazing facts of adaptedness. Life itself is a steady state of enormous general improbability and that which does need an explanation is the fact that organisms and species miraculously manage to stay alive. The answer to this question in respect to bodily structure as well as to behavior always hinges on the provenience of the information contained in and indispensable for the molding of the organism in such a way that it fits its environment and is able to cope with it. In the special case of our baby *Apistogramma* and of our male stickleback we want to know how the former can possess information on the external characteristics of its mother and how the latter can "know" what its rival looks like.

Instead of being faced, as Jensen's proposal would make it seem, with a Herculean task promising only uninteresting results, we have before us a program of practically feasible experiments which simply cannot fail to give interesting results one way or the other. All we have to do is to rear an animal, as perfectly as we can, under circumstances that withhold the particular information which we want to investigate. We need not bother about the innumerable factors which may cause "differences" in behavior as long as we are quite sure that they cannot possibly relay to the organism that particular information which we want to investigate. If our baby *Apistogrammas,* who have never seen an adult female of their species, selectively respond to a certain black and yellow pattern by following it and staying with it (as they would with their real mother) or if our stickleback responds to an object which is red below with the highly specific motor patterns used in fighting a rival, we are justified in asserting that the information which these fishes possess concerning these two objects is fully innate. In other words, the fish's genome must contain the blueprint of a perceptive apparatus which responds selectively to certain combinations and/or configurations of stimuli, and which relays the message "mother present" or "rival male present" to effector organizations equally adapted to dealing with this fact. That much we know, and this is, in my opinion, intensely interesting knowledge. Furthermore, it is knowledge which raises new questions.

First among these are the problems of ontogeny. It is a hackneyed fact that it is never more than just such a blueprint which is inherited and that between this inheritance and its final realization in bodily structures and functions lies the whole process of individual development of which experimental embryology (*Entwicklungsmechanik*) tries to gain causal understanding. This indeed is the truth contained in the statement that organs or behavior patterns must not be called innate. The experimental embryologists, while trying to gain insight into the physiological causality of development, have always acted on the knowledge that any structure which is elaborately adapted to functions and environmental givens to be faced only much later in individual life must be blueprinted in the genome. They also know that adaptive epigenetical regulation can only be expected on the basis of such information that one part of the embryo can receive from another. The ectoderm must have all the information about how to build a neural tube. What the ectoderm can "learn" from the organizer emanating from the chorda is only where to do so. If investigators never put this into words, it was because all this was a matter of course to all of them, but they certainly would not have gained the results they did had they allowed their approach to be narrowed by the attitude exemplified in Jensen's paper.

Any investigator of the ontogeny of animal behavior is

well advised to begin with the time-honored procedure of first searching for whatever may be blueprinted by heredity. This is good strategy and not only for the simple reason that anatomy regularly begins with the investigation of the skeleton; in any system whose function consists of the interaction of many parts, the least changeable are the best point of departure, because their properties appear most often as causes and least often as effects in the interaction that is to be studied. An even better reason is that phylogenetically adapted structure represents, in the ontogeny of behavior, the indispensable prerequisite for guiding modification in the generally improbable direction of adaptedness. We cannot ever hope to understand any process of learning before we have grasped the hereditary teaching mechanism which contains this primary programming.

The emphasis which I am putting on the fact that we know about some indubitably inherited blueprints of motor patterns as well as of receptor organizations must not mislead anyone into thinking that I deem it unnecessary to investigate the ontogeny of behavior. The very opposite is true, particularly as far as receptor patterns (in other words, releasing mechanisms) are concerned. Among these, there is hardly one which, though indubitably based on innate information, is not rendered more selective by additional learning. A flashlike conditioning can take place at the very first function of a releasing mechanism and has to be taken into consideration in order to avoid the danger of mistaking what had actually been learned in the first experiment for innate information in later ones. In order to assess correctly the amount of innate information contained in a releasing mechanism, it is literally necessary to pull it to bits—no pun intended. The method of doing so will be discussed in the chapter on the deprivation experiment (p. 89).

At present, one example is sufficient to illustrate the point. The behavior by which an inexperienced turkey hen responds to her first brood of chicks is dependent on a single phylogenetically adapted receptor mechanism. As Schleidt and Schleidt (1960) have conclusively shown, she treats every moving object within the nest as an enemy, unless it utters the specific note of the chick. A deaf hen invariably kills all of her own progeny immediately after hatching. A hen with normal hearing accepts and mothers any stuffed animal, if it is fitted with a small loudspeaker uttering the correct call notes. Under natural circumstances with a young turkey hen hatching her own young, the one phylogenetically adapted auditory reaction effects so rapid a conditioning of maternal responses to other stimuli emanating from the baby birds that after a few hours the unconditioned stimulus can be dispensed with and the mother is ready to brood the young even when they are silent. Nevertheless, the babies' call notes continue to enhance maternal activities.

The Schleidts did not succeed in finding an artificial substitute for the reinforcement of maternal behavior effected by the turkey chick's call note. Although it was possible with the utmost patience to sufficiently habituate some deaf hens to the presence of little chicks so that these were no longer killed on sight, it proved to be quite impossible to condition any deaf mother's maternal activities so as to respond to the babies. All that could be achieved was that the deaf birds behaved as if there were no chicks present, in spite of the fact that they were in exactly the right phase of their reproductive cycle and so tame that all kinds of manipulation were possible in the attempt to establish a contact between them and their progeny.

My point in telling about these highly interesting results is that they could never have been obtained by using the conceptualizations and methods proposed by Jensen. The operation of withholding specific information underlying adaptedness is only one among literally millions of possible changes "causing differences in behavior," and the chances of finding the effective conditioning mechanism which the Schleidts found by these methods is correspondingly small.

No biologist in his right senses will forget that the blueprint contained in the genome requires innumerable environmental factors in order to be realized in the phenogeny of structures and functions. During his individual growth, the male stickleback may need water of sufficient oxygen content, copepods for food, light, detailed pictures on his retina, and millions of other conditions in order to enable him, as an adult, to respond selectively to the red belly of a rival. Whatever wonders phenogeny may perform, however, it cannot extract from these factors information which simply is not contained in them, namely, the information that a rival is red underneath.

We must be quite clear about what we call a "behavior element." If the Kuenzers' baby *Apistogrammas* responded selectively to a certain color configuration which roughly but sufficiently corresponds to the coloration of the mother they have to follow, we certainly must assume a nervous mechanism which, among innumerable other possible visual stimulus situations, selects this particular one and connects it with effector patterns phylogenetically adapted to keep the little family together. This mechanism is, of course, dependent on a structure built during ontogeny on the basis of a genetical blueprint. The whole function of specifically responding to the maternal color patterns involves functions other than that of this particular mechanism. Many of the processes which take place on the way from sensory stimulation to the response are less specific than the latter. Not only the processes of visual stimulation in the retina, but much more highly integrated functions, like those of perception (including depth perception, color constancy, and so on), are used in very many other responses and/or activities of the animal. Among these functions there may be some that require ontogenetically acquired information for their full development. Even the function of retinal elements requires "practice"; it is well known that retinal elements are subject to atrophy if not sufficiently used. The faculty of point discrimination, probably performed by the ganglion retinae, is lost to a large extent if not practiced, even if diffused light and unfocused images do impinge on the retina. If a person's vision is impaired for a long period by purely optical deficiencies of the eye, point discrimination remains seriously damaged even after correction of the optical apparatus, a state of affairs termed *amblyopsia ex anopsia* by oculists.

It is a matter of taste whether or not one choses to call it learning when an activity is necessary to prevent atrophy and disintegration of a physiological mechanism, but it can be regarded as adaptive modification and it may well involve ontogenic acquisition of information. Much the same is true of the effector side of the reaction. Orientation mechanisms, motor patterns, etc. functioning in a stickleback fighting a rival or in an *Apistogramma* baby following its mother may also be elements of behavior that occur in other contexts as well. Among them, too, there may be some that need an inflow of individually acquired information for their full functional development.

These modifications adapting the less specific elements of a more specific response are indispensable requisites for the proper functioning of the latter. They must, therefore, never be forgotten or overlooked in our attempts to analyze this function. On principle, however, they are no obstacle to the solution of our fundamental question concerning the provenience of the information underlying each point of adaptedness in behavior. Nor do they represent a very serious source of error quantitatively. What may be overlooked is the effect of a little sensory adaptation making an afferent process a little more selective, or that of a little practice smoothing out a motor coordination. But there is little danger, with circumspect experimentation (see page 90) and with an experimenter knowing its pitfalls, that any process of true learning, particularly classical conditioning, might pass unnoticed.

It must not be thought that the learned prerequisites for the proper functioning of innate information are only made up of the primitive types of adaptive modification of behavior. True conditioning does figure among them. A good example of this is furnished by the copulatory response of some geese. Greylags (*Anser anser*), Greater snow geese (*Anser hyperboreus atlanticus*), and probably many others possess perfectly good phylogenetic information about how a fellow member of the species behaves when inviting copulation, but they have to learn what a fellow member of the species looks like. The genetic information may be verbalized as follows: copulate with a conspecific who is lying low in the water and is stretched out along its surface. Hand-reared greylags, usually born in the first days of May, usually do not meet their human foster parents swimming in the water earlier than June (in our climate). They then regularly attempt to tread them. As a human swimmer is much longer than any goose and also is lying much lower in the water, he represents a supernormal object for the copulatory response of all those geese which regard him as a conspecific. Even in very young fledgling geese and also in females, which normally never show any copulatory movements, attempts to tread can reliably be released by this supernormal stimulation. These phenomena definitely are not a consequence of imprinting. The geese behaving thus remain otherwise quite normal in respect to their sexual objects and do not persist in trying to copulate with humans. Nor do they ever attempt copulation with any other randomly chosen object that happens to be elongated and lying low in the water. These configurational properties are effective as key stimuli only if they pertain to an object which the goslings have learned to know as a fellow member of their species, and this learning is achieved by all the many conditioning processes which attach the gosling to its parents and its siblings.

Much ontogenetically acquired adaptive modification of behavior, may be involved in a functional whole, such as fighting a rival in the stickleback, following the mother in *Apistogramma,* or copulating with a conspecific in geese, however it does not affect either the correctness or the justification of our statement that certain parts of the information tion which underlie the adaptedness of the whole and which can be ascertained by the deprivation experiment are indeed innate.

It is obviously this information alone to which we have a right to apply the term innate. We cannot, however, think of any way for this information to express itself in adapted behavior other than by the function of a neural structure. It is the distinctive property of this structure to select, from among innumerable other possible stimulus situations, the one which specifically elicits the response. This property definitely is a character of the species. The neural mechanism is an organ and not a character. A character in *The Oxford English Dictionary* is a "distinctive mark."

When some geneticists and, following them, many modern ethologists contend that characters must never be called innate, I have a suspicion that they are confounding the concepts of a character and of an organ. The latter, of course, really must not be called innate nor, indeed, must a behavior pattern. The formulation that it is not characters but differences between characters which may be described as innate is, in my opinion, an unsuccessful attempt to arrive at an operational definition. If we rear two or more organisms under identical conditions, any differences shown by them may be regarded as caused by differences in the genome. Theoretically this is all right, but I doubt whether the definition can be put into operation. If we rear a number of larval fish in the same tank under "identical conditions," for instance with very few larval crustacea present, the few fish which happen to catch these few nauplii will grow much faster than those that don't. How can we prevent, on principle, analogous sources of error? It seems to me that the opposite formulation is at least as workable: calling innate the similarities of characters developing under dissimilar rearing conditions. If we observe that all mallard drakes—and many other male dabbling ducks—whether reared in the wild or in captivity, by their own mothers or by a human keeper, under good or under highly unnatural conditions, perform the grunt-whistle in very nearly the same way, the breadth of variability being almost negligible in the confines of one species, our assertion that this similarity is innate, that is, based on genetical information, has at the very least the same likelihood of being correct as the opposite one, that dissimilarities in identically reared organisms are innate. It is, indeed, the similarity of individuals that tells the taxonomist what a species really is, and he was never mistaken in applying this indubitably genetical concept even in Linné's time.

Our question, "Whence does the organism derive the information underlying all adaptedness of behavior?" leads directly to more special and to more general questions. What rules ontogeny, in bodily as well as in behavioral development, is obviously the hereditary blueprint contained in the genome and not the environmental circumstances indispensable to its realization. It is not the bricks and the mortar which rule the building of a cathedral but a plan which has been conceived by an architect and which, of course, also depends on the solid causality of bricks and mortar for its realization. This plan must allow for a certain amount of adaptation that may become necessary during building; the soil may be looser on one side, necessitating compensatory strengthening of the fundaments. The phylogenetically adapted blueprint of the whole may rely on subordinate parts adaptively modifying each other. The prospective chorda exerts a very specific adaptively modifying effect on the neighboring ectoderm, causing it to form a neural tube in exactly the right place. The adaptive modification effected by one part on the other may even take the form of true learning. The gosling "knows innately" that it should copulate with a fellow member of the species stretching out low in the water, but it has to learn what a fellow member of the species is. Any such adaptive regulations, however, presuppose at the very least as much information contained in the genetical blueprint as any elements of little or no modifiability do. In other words, the apparatus which makes adaptive modifiability possible is genetically blueprinted itself, and it is in a very complicated form, particularly if it allows so much scope to regulations as we find in embryogeny.

Because all the causal chains of development begin with the hereditary information contained in the genome (and the plasma) of the egg, our first question concerning the ontogeny of an organism and its behavior is: "What is blueprinted in its genome?" The second is: "What are the causal chains which begin at the blueprints given in the genome and which end up, by devious and often highly regulative routes, by producing adapted structure ready to function?" We are fully aware that the processes of growth can be separated from those of behavior only by an injunctive definition permitting intermediates. We think, however, that the process of individual learning is sufficiently distinct from other processes of adaptive modification, and particularly from those regulating structural growth, that we are justified in leaving, at least for the time being, to the care of the experimental embryologists all those questions which are concerned with the chains of physiological causation leading from the genome to the development of such neurosensory structures. These structures, like the releasing mechanism of the stickleback's rival fighting or like that of the turkey hen's response to the chick's call, demonstrably owe their specific adaptedness to information acquired in phylogeny and stored in genes.

Not being experimental embryologists but students of behavior, we begin our query, not at the beginning of the growth, but at the beginning of the function of such innate mechanisms. The modern English-speaking ethologists are generally agreed that the concept of the "innate" is valuable and valid only if defined as "not caused by modification." This cautious definition, although of course quite true as far as it goes, states only the less important side of the problem. The important one is that the phylogenetically adapted structures and their functions are what effects all adaptive modification. In regard to behavior, the innate is not only what is not learned, but what must be in existence before all individual learning in order to make learning possible. Thus, consciously paraphrasing Kant's definition of the A Priori, we might define our concept of the innate.

This definition is truly operational, as can be seen in the Schleidts's and many other people's investigations. On the other hand, I defy anybody to put into practical operation the allegedly operational definition proposed by Jensen. Whoever tries it, will, with an overwhelming probability, die of old age before reaching publishable results. Hence the acceptance of this definition must discourage any investigation of the ontogeny of behavior.

Obviously it is the best strategy of research if, in our attempt to analyze the ontogeny of behavior, we first concentrate on the question: "What are the teaching mechanisms?" and second on the question: "What do they teach the animal?" In asking these questions, we may seem to neglect, at least for the time being, those "behavioral differences" which are "neither innate (selected) nor learned (trained) but something else (produced by other operations)." When we rear, in our experiments, an organism under circumstances which are calculated to withhold from it some specific information, we do our very best to avoid that kind of difference in behavior. In other words, we try to produce an individual whose genetical blueprints have been realized unscathed in the course of healthy phenogeny. Should we fail in this, we would incur the danger of mistaking some defects in our subject's behavior for the consequences of information withheld, while they really are the pathological results of stunted growth. This danger is very great if the investigator is not aware of it, if he does not know the system of actions of his subjects inside out, and if he does not possess what is called the "clinical eye." Otherwise, the danger is almost negligible.

Of course, the behavior of an otherwise perfectly healthy mandarin drake (*Aix galericulata L.*) whose sexual responses are fixated by the process of imprinting on mallard ducks (*Anas platyrhynchos L.*) is also indubitably pathological, particularly from the viewpoint of survival, as the bird is rendered permanently unable to reproduce. But not only do we know what behavioral system we have damaged in this mandarin by the intentional misinformation furnished to it at a critical stage of its ontogeny, but even if we did not know the previous history of the individual, we could still, with great certainty deduce it from the symptoms (pp. 90, 91).

In the present chapter, criticizing the assumption of a general and diffuse modifiability of phylogenetically adapted behavior mechanisms, I need only say that the enormous symptomatic difference between behavior defects caused by withholding information and defects "produced by other operations" is in itself extremely strong and convincing evidence against that assumption.

Finally, I want to discuss a speculative but, to me, rather convincing argument against assuming a diffuse modifiability of phylogenetically adapted behavior through learning. I have already said (p.12) that with the increasing complication of an adapted system, the probability decreases that random change in any of its parts may produce anything but disadaptation. The power of the argument illustrated by the sports car is multiplied greatly when we consider really complicated neural systems, for instance, those performing real computations. Phylogenetic adaptation has created mechanisms of such subtle complication here that even a really brilliant physiological cyberneticist is barely able to gain a tolerably complete insight into their workings. After carefully reading and rereading Mittelstaedt's paper (1957) on the complex feedback mechanism enabling a mantid to aim a precise stroke at its prey, my own understanding of that mechanism is insufficient to permit sensible suggestions for its improvement. So I find it difficult to believe that the insect should be superior to myself in that respect, unless, of course, it possesses special built-in calibrating or adjusting mechanisms. I argue that this type of complicated neural mechanism must be highly refractory to random change by individual modification. The complexity and precision with which the processes of evolution have endowed these mechanisms would be destroyed immediately if individual modification by learning were allowed to tamper with them.

Even in man, computing mechanisms of this kind, particularly those of perception, are built in such a way as not to let learning "enter into" them. Although these mechanisms very often perform functions so closely analogous to rational operations that Brunswik (1957) termed them "ratio-morphous" processes, they stubbornly refuse to have anything to do with rational processes, least of all to let themselves be influenced by learning. As von Holst (1955, 1957) has shown, all so-called optical illusions, with very few exceptions, can be regarded as the results of ratio-morph operations which are caused to miscarry by the introduction of one erroneous "premise." From these, a perfectly logical computation draws a false conclusion. These mechanisms are not only inaccessible to self-observation but also refractory to learning.

If we consider learning as a specific function achieving a definite survival value, it appears as an entirely unfounded assumption that learning must necessarily "enter into" all other neurophysiological processes determining behavior. It is, however, by no means this theoretical consideration alone which causes us to reject that assumption; all experimental and observational evidence points that way. Not once has diffuse modifiability been demonstrated by experimentally changing arbitrarily chosen elements of phylogenetically adapted behavior mechanisms. On the other hand, innumerable observations and experiments tend to show that modifiability occurs, if at all, only in those preformed places where built-in learning mechanisms are phylogenetically programmed to perform just that function. How specifically these mechanisms are differentiated for one particular function is borne out by the fact that they are very often quite unable to modify any but one strictly determined system of behavior mechanisms. Honey bees can learn to use irregular forms, like those of trees or rocks, as landmarks by which to steer a course to and from the hive; but, they cannot, even by the most subtle conditioning technique, be taught to use these same forms as positive or negative signals indicating the presence or absence of food in a tray as von Frisch (1914) has shown. As signals for food, bees can distinguish different forms only if they are geometrically regular, preferably radially symmetrical (Hertz, 1937).

In other words, the old and allegedly naïve theory of an

"intercalation" of phylogenetically adapted and of individually modifiable behavior mechanisms, far from having been refuted by new facts, has proved to agree with them in a quite surprising manner. Scientists working on entirely different problems, such as the selectivity of innate releasing mechanisms, bird navigation, feedback mechanisms of aiming, circadian rhythms, etc. have one and all found that if modifiability existed at all, it was restricted to one particular link in the chain of neural processes determining behavior. To demonstrate my point, I shall discuss a few examples of "circumscript" learning. As we know, practically nothing about the physiological processes underlying the different types of learning, these examples can be classified only from a functional point of view.

SOME SPECIFIC FUNCTIONS OF LEARNING

As Thorpe (1956) has pointed out, the simplest and phylogenetically oldest way of adaptively modifying behavior is the waning and, finally, the cessation of the response to a biologically irrelevant stimulus constantly repeated throughout a longer period. Thorpe has defined this phenomenon, which is generally called habituation, as "the relatively persistent waning of a response as a result of repeated stimulation which is not followed by any kind of reinforcement." While this definition justly stresses the size order of duration as a point discriminating habituation from fatigue and/or sensory adaptation, it implies, on the other hand, a close physiological relationship between habituation and the conditioned extinction of a response. I use this rather unorthodox phraseology advisedly because I want to emphasize that not only conditioned responses but also "unconditioned" ones can be extinguished by a conditioning process. It is an unconditioned response which makes the newly metamorphosed toad (*Bufo vulgaris L.*) snap at any moving object of suitable size, and it is a process of true conditioning, by the extinguishing effect of the impossibility to perform the consummatory act of eating, that the baby toad learns not to snap at moving grass blades and other inedible objects (Eibl-Eibesfeldt, 1951). This case of indubitable habituation conforms exactly with Thorpe's definition. There are other cases of habituation, however, in which reinforcement definitely has no influence on the decrement of the response, as has been conclusively shown by Hinde (1960) in his work on the mobbing response of the chaffinch (*Fringilla coelebs L.*). Habituation occurs, moreover, in organisms like protozoans and coelenterates, in which no true conditioning could ever be demonstrated.

I would therefore propose to leave out, in our definition of habituation, the words "which is not followed by reinforcement" and substitute for them a functional property of habituation. Again it was Thorpe who emphasized that the main survival function hinges on the fact that the response is only suppressed in answer to that particular stimulus situation which has engendered habituation. It certainly is very important that habituation spares the organism the necessity of continually responding to stimuli which, although primarily eliciting a reaction, are of no biological relevance. A bird whose escape responses are released primarily by anything that moves, as well as by anything new, would never come to rest at all, if it were not for its habituation to constantly recurring stimuli of this kind. These advantages of habituation would be counteracted by severe drawbacks, if the response to all other stimuli which also elicit it would suffer a decrement as well. The really surprising function of habituation, and the one in which its survival value lies, is the elimination of the organism's response to often recurring, biologically irrelevant stimuli without impairment of its reaction to others. A hydra is thus able to disregard water turbulence, while its responses to all other stimuli eliciting contraction remain as finely triggered as they ever were. The bird ceases to be frightened by moving branches and falling leaves without becoming, even in the slightest degree, less responsive to other movements that might spell danger.

I suggest, as a definition of habituation, a combination of the two given by Thorpe: habituation is the relatively persistent waning of a response to certain stimuli caused by their repeated impinging and not affecting the threshold of the response in respect to any other than the habituated stimulus situation. This characterizes well enough what we know about the survival value of the process, and, as long as we know as little as we do about its physiological causation, it is advisable to confine ourselves to a purely functional definition which includes those cases of habituation in which conditioning takes a part as well as those in which it does not.

The following observations illustrate the subtle selectivity of response achieved by habituation. In the greylag, the innate mechanism releasing the escape response to the one flying predator endangering this species (the white-tailed eagle) is comparatively simple and unselective. The response is released, as Tinbergen has shown in experiments conducted in Altenberg in 1937, by any object silhouetted against the sky and gliding along slowly—as measured in multiples of its own length—and smoothly without any additional quick fluttering movements (which immediately extinguish the goose's reaction). Very slow, measured wing beats seem to increase it; however, this has not been experimentally confirmed. As a consequence of the small number and simplicity of these key stimuli, this escape response can be released by many objects other than white-tailed eagles. A dark feather slowly floating along on a quiet breeze, a pigeon or a jackdaw gliding against a head wind which slows its progress, or a buzzard or a big plane high in the sky are all primarily "models" eliciting the response just as well as does an eagle. With repetition of any of these stimulations, however, the eagle-response of the geese wanes very quickly by virtue of a habituation which is extremely specific to the stimulation that has effected it. The threshold of the response to a buzzard is not affected by the habituation to planes nor does the waning of the reaction to buzzards and planes, both frequent in our region, influence the one elicited by the much rarer appearance of a grey heron. All these habituations together do not weaken, in any way, the response to a white-tailed eagle, would one ever put in an appearance.

A phylogenetically adapted releasing mechanism which has never as yet been experimentally analyzed but which is certainly more complicated than the one just discussed elicits the intense "mobbing" response which geese and other Anatidae direct at a fox stalking along the water's edge. When we moved our waterfowl from Buldern into an unenclosed area in which foxes abound near Ess-See, I was afraid that the birds' relative fearlessness of dogs and particularly of my Chow-Alsation crosses (which, from the point of view of the corresponding releasing mechanism, are perfectly supernormal foxes) might impair their wariness of these very dangerous predators. The dramatic manner in which this danger was avoided by the specificity of habituation to highly complex stimulus situations was actually what first made me realize its tremendous survival value.

This also alerted me to the existence of an entirely unsolved problem. Decrement of response to a repeated stimulus situation is a very general occurrence, and I do not regard it a safe assumption that it is more marked in more general, less selective releasing mechanisms than it is in highly specific and selective responses that are waning quickly and irreversibly, as Hinde has shown in the mobbing response of the chaffinch. Attempts to prevent this decrement by applying reinforcement were entirely without avail, and this fact faces us with the question whether, in natural wild surroundings, a chaffinch's response to owls is as quickly exhausted. This is hard to believe considering the high selectivity and the obvious survival value of the phylogenetically adapted response. In such cases, the obvious unadaptivity of the reaction decrement definitely makes a misnomer of the term "adaptation" as it is used by many physiologists dealing with the senses. In young greylag geese, the escape response elicited by an imitation of the parent's warning call fades just as quickly and as irreversibly as the mobbing of an owl model in the chaffinch. No decrement at all, however, is observable in the goslings' response to the warning of their real parents, which makes me suspect that possibly the quick waning might occur only in a response to subnormal stimulation of a specific response. Another interesting possibility, to which Hinde was the first to draw attention, is that an abnormally quick waning of response is caused by the great uniformity of circumstances accompanying the presentation of stimuli in the experimental setup, while, in wild life, no response is ever elicited even twice under exactly identical conditions.

It has been the repeated experience of ethologists experimenting with dummies of all sorts that uniformity of place or of movement, regularity of rhythmically repeated sound, etc. cause a very fast waning of response. My own studies on the waning of the goslings' following response as well as of their reaction to the alarm note of the parents, Kühme's investigation (1962) of the following response of young cichlid fish, and the Schleidts' research (1961) on the turkey's escape reaction released by flying predators were in perfect agreement on this point.

It is, at the present state of our knowledge, a matter of pure speculation what physiological mechanisms may cause the function of habituation, whose very definite survival value was discussed in this chapter. It may be physiologically akin to fatigue. It may even have been evolved from fatigue phenomena which became more and more specific in respect to the stimulus situation that caused them, but its present survival value lies in preventing certain effects of fatigue which might be detrimental to survival. One assertion, however, which we can make with some certainty is that very different physiological processes may take part in achieving this survival value. In the habituation of geese to individual dogs, processes of *Gestalt* perception indubitably play an important role, while in the habituation of hydra to turbulence of water they certainly do not. If I have chosen many examples in which the specific function of habituation is achieved without the participation of conditioning, I have done so only in order to show that this does occur, and not to deny the irrefutable fact that the same function can be, and very frequently is, performed by negative conditioning, as in the case of Eibl-Eibesfeldt's freshly metamorphosed toads. To stress the fact that habituation is not, on principle, identical with negative conditioning seemed important to me because the same is true of another specific function of learning which is the subject of the following chapter.

INCREASE OF SELECTIVITY IN UNLEARNED, STIMULUS-SPECIFIC RESPONSES

In common parlance, the expression "getting habituated" or "accustomed to"' a stimulus situation can also signify the very opposite of the process that has been discussed. In the latter, an additional combination of stimuli is woven into an inseparable whole with the key stimuli which, innately and originally, are alone effective in releasing the response; but although, in the process of habituation, the newly acquired stimulus situation after numerous repetitions extinguishes the innate response to the key stimuli contained in it, the process under discussion here extinguishes the response to all other stimulus situations, even though they also contain all the releasing key stimuli. As in the one discussed in Chapter 4, this process can be accomplished without the participation of conditioning.

Immediately after hatching, a greylag goose, will react to any object which in answer to its distress signals (lost piping) utters rhythmical sounds over a wide range of pitch. After the process of imprinting (p. 55), on leaving the nest it will follow its mother or the dummy which has supplied sufficient stimulation for imprinting. At that time it can distinguish its mother from a man or from a dummy, but it is quite ready to follow any other goose. A few days later, it will follow its mother only and recognize her call at considerable distance, responding to it in preference to identical stimuli emanating from a much nearer mother leading goslings of its own age group. There is no evidence at all that conditioned extinction of the response takes part in these proceedings. Goslings which, during the first hours or days spent out of the nest, have lost their own mother and, after the most gruelling experiences, have found her again, are not less likely to lose her again than those which have not undergone this punishment, but significantly more prone to lose her. In very many species of birds, the response of the parents to their offspring undergoes a similar increase in selectivity, also without any negative conditioning to other situations. Immediately after hatching their own eggs, they will accept any young of approximately equal age, but later they recognize their own chicks individually and refuse to take care of strangers. Under the abnormally uniform conditions of captivity, this increase of reactional selectivity can lead to the weirdest form of "pedantry." Old cage birds who have eaten from the same food tray for many years will often persistently refuse to eat from another tray or from the ground. In all these cases, the phylogenetically adapted releasing mechanism is supplemented by a great number of additional conditions which must be fulfilled in order to make it respond. This increase in selectivity takes place without any negative conditioning, as was well known to the classic investigators of the conditioned response.

A process closely related to the one just described is what we call imprinting. It is an old rule of thumb that no phylogenetically adapted releasing mechanism is more selective than is necessary to prevent the response from being released by another than the biologically adequate object with greater frequency than is compatible with the survival of the species. In cases in which individual experience gets an opportunity to increase the selectivity of a releasing mechanism, the latter can "afford" to rely on very simple and generalized innate information, particularly if the general biological situation guarantees that the right kind of experience sets in at the right moment. In such higher animals that are either social in habit or that perform parental care, responses to the conspecific are often dependent on simple, and therefore unselective, releasing mechanisms, while the social environment of the young animal dependably supplies the conditions for additional learning, and this learning rapidly builds up a high degree of selectiveness in the mechanisms releasing social responses.

In some cases, a very strict limitation of this learning process to a definite period in the individual's life compensates for the simplicity and lack of selectivity of the releasing mechanism to be improved by learning. In a greylag goose, the mechanism releasing the following response of the newly hatched gosling consists of the receptor organizations corresponding to only three configurations acting as key stimuli: (*a*) noise of a widely variable pitch occurring in "answer to" the gosling's "lost piping," (*b*) movement of the object away from the gosling, and (*c*) rhythmical sounds emanating from object. The gosling crawling out from under an artificial brooder after drying will sit in front of it and after a short time utter its "lost piping." If the brooder is fitted with a small loudspeaker or a simple buzzer and responds to the "lost piping" by a noise, the gosling becomes silent and looks up at the brooder. If this is continued until the bird is able to leave the nest, it will follow the brooder closely the first time the brooder is moved away from its place.

After this process, which of course takes place in the same way between a gosling and its real mother, the following response is fixated on the object in question. This fixation is irreversible, at least insofar as any substitution of a different object results in a decrement of the response's intensity. Although indubitably coming under our definition of learning, this process is distinct from our accepted concept of conditioning in several interesting ways. The first, and probably most important, is that a response becomes conditioned to a highly specific stimulus situation without actually having been released. This is particularly striking in many known cases of sexual "imprinting." The second, and still unintelligible, point is the generic quality of this fixation of the object of a certain response. In Schutz's (1963, 1964) experiments on sexual imprinting, his subjects, mostly male mallards, were kept for the first seven to eight weeks of their lives in the exclusive company of mothers and/or siblings of another duck species. After that, they were liberated on the Ess-See. Invariably they flocked normally with other mallards until the awakening of their sexual responses. These, in the majority of cases, were directed at members of the species with which they had been reared without any preference for individual animals.

Similarly, a gosling's following responses become imprinted "to geese" or "to humans," but it is still an intriguing puzzle which particular stimulus situation represents each of these concepts. The human-imprinted gosling will unequivocally refuse to follow a goose instead of a human, but it will not differentiate between a petite, slender young girl and a big old man with a beard.

During the first two days after leaving the nest, the gosling follows the parent about and there is a second tremendous increase in the selectivity of its following response. By that time, it has learned to recognize the parents personally and will never mistake another pair of geese for them, even if they are leading a flock of babies of the same age.

Probably the two processes, which, in two steps, increase the selectivity of the mechanism causing young goslings to follow the parents, are physiologically different from each other. The two processes can occur independently of each other, as shown by the lack of one or the other in certain species. In muscovy ducks, for instance, it is easy to imprint the sexual responses of the drake on any other species of Anatidae by using members of this other species as foster parents. Drakes thus treated will copulate with any bird of the species they have been imprinted on, but they never react to their partner individually for the simple reason that muscovies do not do so to their conspecifics either. Conversely, the females of mallard and European teal cannot, as far as we know, be sexually imprinted on other species, because their sexual responses are too strongly dependent on the key stimuli emanating from the conspecific male's nuptial plumage.

The examples hitherto cited, in order to illustrate the increase of selectivity in responses to a specific and repeatedly impinging stimulus situation, were chosen intentionally to show that this process can take place without the participation of true conditioning involving reinforcements. Of course, an originally unselective releasing mechanism can also be made more selective by true conditioning, in other words, by reinforcing the response to that particular situation and by extinguishing those to all others. This is what happens in the case of the freshly metamorphosed common toad.

CALIBRATION OF AIMING MECHANISMS, ADJUSTMENT OF COMPUTERS AND SETTING OF "INTERNAL CLOCKS"

A rather peculiar kind of circumscript modifiability is "built into" some highly differentiated, phylogenetically adapted mechanisms, such as those of aiming, of complicated computers which calculate the points of the compass on the basis of the information given by the moving sun, or of the internal clocks whose function is a prerequisite to navigatory computers. It is at least doubtful whether any of these modificatory processes have anything to do with the establishing of conditioned responses, although, in the sense of our purposely wide definition, they indubitably represent "learning."

As Mittelstaedt has shown in the case of prey catching in mantids, aiming mechanisms can consist of very complicated regulatory cycles in which the information derived from the shift of the retinal image as well as from proprioceptors is fed back into the activity of the motor organization in such a manner as to achieve an optimal compromise between the conflicting exigencies of quickness and precision of aiming. In this feedback circle, a link effecting calibration is not a logical necessity, but its existence can perhaps be inferred from the fact that in some cases the consequences of operations which disturb the regulatory cycle are compensated, at least partially, after a certain time. Newly hatched chickens which Hess (1956) fitted with prism glasses causing a lateral shift of both retinal images in the same direction never learned to compensate this apparent displacement of the object at which they were aiming. Symmetrical prisms which made the goal appear to be nearer than it was caused the chicks to peck short at first, but they learned gradually to correct this. Proprioceptors seemed to participate in this adaptation, because it worked only if the food at which the chicks pecked was lying in the plane on which the birds were standing. The birds persistently pecked short of food which appeared to be floating in mid-air because it was presented on the ends of thin wires.

In ants which, as has been known for some time, are able to steer a course to and from the nest by taking their bearings from the sun, Jander (1957) has shown that learning is built into one strictly defined place of the highly complicated regulating system. When running from the nest, the insect is under the permanent and unchanging influence of a positive phototaxis and/or a negative geotaxis which, as is often the case in insects, are mutually interchangeable. When the "mood of going out" changes into that of "homing," the sign of these taxes changes simultaneously—or rather this change of sign is what causes the ant to turn back. The mechanism which enables the insect to do much more than blindly obey these simple taxes and which endows it with the ability to steer a "voluntarily" chosen course or run along a learned pathway from the nest to a source of food and back again is an additional and rather complex central nervous mechanism which, like the taxes, also generates impulses to turn and superimposes the latter on those effected by the constant taxes. Unlike the latter, this "order to turn" (*Drehkommando*) changes in strength with a number of factors. Its strength determines the magnitude of the angle at which the insect deviates from the course dictated by the simple taxes. The magnitude of this angle is dependent on two mechanisms. One computes the points of the compass from information supplied by the moving sun; the other is the one "into which learning enters"; that is, the ant is able to integrate the universal pattern of optical stimuli which is characteristic of the precincts of the nest in a manner that determines the "center of gravity" of illumination. The ant uses this as a landmark by which to steer a course by keeping a constant angle to it. Detours are mastered by another mechanism of integration which is able to record and "remember" the sequence in which, on the way from and toward the nest, the landmarks follow each other (integration of stimulus sequence).

After investigating the complex regulative interaction within this orienting mechanism, Jander represented it in the form of a cybernetic diagram which corresponds to facts at the very least insofar as every "box" of the diagram corresponds to a real function which was successfully isolated experimentally. It can be shown that the results computed by the mechanism compensating the movements of the sun as well as the learning processes both have a direct influence on the "order to turn."

One case, which in a particularly striking manner illustrates the specific phylogenetic adaptedness of learning processes and the way in which they are "built into" complicated neural systems in a strictly defined place, concerns the orientation by the sun in some sunfish (*Centrarchidae*) and cichlids (*Cichlidae*). Braemer and Schwassmann (in preparation) found an important difference between fish of these two groups. Sunfish occur exclusively in the northern hemisphere. When reared in artificial light, they are able, at their first encounter with real sunlight, to take their bearings from a sun moving in the way it does in their habitat, from left to right. Confronted with a sun moving in the opposite direction, as it does south of the equator, they persist in computing the points of the compass on the "assumption" that the sun moves as it does in the North. Quite unlike sunfish, the young of the cichlid, *Aequidens portalegrensis,* reared in the same manner are able to take their bearings equally well from a northern or from a southern sun, depending only on which way the sun was moving when they first saw it. As the distribution of the species reaches north as well as south of the equator, both faculties of computing are obviously necessary to it.

Another comparatively simple modificatory mechanism built into rather complicated phylogenetically adapted systems is the effect of the "time giver" (*Zeitgeber*) in "setting" the "internal clock" of so many circadian rhythms. As Aschoff (1962) has shown, none of the "chronometers," whose function is the prerequisite of the mechanisms computing the points of the compass from the direction in which the sun is standing at a given moment, is nearly as exact as any average clock of human making. They are all either fast or slow by a very considerable amount, and they have to be "set" anew every day in order to preserve the organism's information concerning time. This "setting" of the clock is accomplished by a modifying response to an outer stimulus directly or indirectly timed by astronomical processes, mostly by changes of light or temperature. This inexactitude of the clock and the necessity of its daily resetting is so much a matter of course to the investigators of rhythmical processes that in the rare case when they do encounter a clock which, under strict deprivation of time-giving stimuli, still runs exactly twenty-four hours a day, they tend to conclude, not that the clock is exact, but rather that the experimental setup is not, in other words, that a hidden time giver is at work.

In all investigated cases, the speed of the clock mechanism when running free was not changed by time givers effecting a daily setting of the clock. In other words, there is no case on record in which the modification effected by the time giver did what the human clockmaker would have done by adjusting the pendulum or the spring of the balance. The effect was invariably confined to what the layman does in adjusting the hands of his watch, which is exactly what one would expect on the basis of the considerations expounded on pp. 45, 46.

FUNCTIONAL INTEGRATION OF MOTOR PATTERNS THROUGH LEARNING

It is highly characteristic of the ontogeny of behavior in higher animals that a phylogenetically adapted motor pattern makes its first appearance in tolerably complete, or at least clearly recognizable, form but in a biologically inadequate situation. A puppy performs the shaking movement, adapted to the killing of prey, with its master's slippers for an object or the motor pattern of burying food remnants on the parqueted floor in the corner of a room. In his classic paper on appetites and aversions, Craig (1918) has demonstrated in a masterly fashion that the phylogenetically adapted mechanism of the consummatory act itself is so constructed as to impart to the subject the "knowledge" of the environmental situation in which to discharge the motor pattern in question. Craig illustrated the principle by the graphic description of a young dove learning "to obtain the stimulation adequate for a complete consummatory reaction, and thus to satisfy its own appetites." The movements of the consummatory act itself are, as Craig repeatedly points out, wholly innate. Craig held them to be true chain reflexes. He does not hesitate, nevertheless, to attribute to the bird highly pleasurable subjective experiences correlated with the performance of the consummatory act. When a young dove, sexually mature but inexperienced, is supplied with a nest for the first time, it does not recognize it at sight:

> But sooner or later he tries it, as he has tried all other places, for nest-calling, and in such trial the nest evidently

gives him a strong and satisfactory stimulation (the appeted stimulus) which no other situation has given him. In the nest, his attitude becomes extreme: he abandons himself to an orgy of nest-calling (complete consummatory action), turning now this way and now that in the hollow, palpating the straw with his feet, wings, breast, neck and beak, and rioting in a wealth of new, luxurious stimuli.

Much as a stickler for objective terminology might have to criticize in this representation, the fact remains that from now on this young dove is thus effectively conditioned to the nest.

The learning processes which Eibl-Eibesfeldt (1955–1963) investigated in rodents and small carnivora illustrate the same principle. Laboratory rats reared in such a manner that they could not gain any experience in the handling of solid objects and even deprived of their own tails (after it had been observed that they performed nest-building activities with this appendage for a substitute object) immediately displayed a number of complete motor patterns of nest building when Eibl-Eibesfeldt supplied them with suitable material. Long sequences of movements, such as running out, grabbing material, running back, and depositing it or arranging it in a circular heap around the prospective nest center, "upholstering" the inner wall by patting soft material smooth by alternating movements of the front paws, or manufacturing soft material by splitting coarse strands longitudinally, etc., were indistinguishable from those of normally experienced control animals, even in the analysis of slow-motion films. Yet the following learning processes proved to be indispensable to integrate these motor patterns into a functional whole.

If the cage was entirely devoid of structure which, by

eliciting a phylogenetically adapted response, marked a preferable nesting locality, the rat had to decide by learning on the spot in which to deposit the nesting material it was carrying. Individuals which, in spite of the complete lack of structure within their cages, had formed a habit of sleeping in a certain corner started to build there at once when offered material without any preliminary trial and error. A tin screen about two inches square was sufficient to direct the inexperienced rat's very first acts of building to the cover it offered. The heaping-up movements and the upholstering movements were often observed when the rat had only carried in two or three paper strips which were lying flat on the ground. The movement was then performed, in perfectly normal coordination, an inch or so above the nest material without ever touching it. This never happens in a normally experienced rat. It will not adjust the movement to the height attained by the accumulated nesting material, but it will not perform it at all until the heap has attained the elevation necessary to bring it within the range of the fixed motor pattern.

Obviously it is the reinforcing effect of the consummatory situation with all the proprio- and exteroceptor reafferences which arise under adequate environmental conditions which, in their joint effect, teach the animal what to do or, by their disappointing absence, what not to do, and particularly, in what sequence to use otherwise unchangeable fixed motor patterns. The information concerning the biologically "right" environmental situation is, in this case, not only contained in the organization of receptor patterns alone, but also essentially in the fixed motor pattern itself which produces the reinforcing reafferences only if very definite environmental conditions are fulfilled. I do not see how a complete study of the learning functions of a species can be accomplished without an investigation of its fixed motor patterns and the reinforcing consummatory situations pertaining to each of them.

This specific function of learning is different from those described in the preceding chapters in one important point: it is dependent on a gradient of more effectively and less effectively reinforcing stimulus situations. Purposive psychologists, particularly Tolman (1932), have always correctly emphasized that learning of this type can take place exclusively in the context of "purposive," or, as we should call it, appetitive behavior. Whether the latter is directed at the releasing of an instinctive movement or at a state of quiescence, in other words, whether it is an appetite or an aversion in Craig's terminology, is not essential for the present issue, because the conditioning effect or "reward" is different from that of "punishment" only in sign. The information on the adaptive optimal consummatory stimulus situation itself is, of course, phylogenetically acquired. The mechanism which thus directs the learning process toward survival value is situated within the selective receptor processes constituting the afferent side of the "unconditioned reflex" which, as is known through the work of Pavlov, is the indispensable basis for the formation of any conditioned response.

Wherever trial and error followed by subsequent adaptive modification of behavior is observed, the assumption of Pavlovian conditioning is justified. Conditioned responses of the second and third order can be attached, forming chains of what purposivists would term means-end relationships. Learning processes of this type can thus accomplish survival functions of almost unlimited complexity.

THE FORMATION OF PATH HABITS

Possibly the path habits found in many higher animals are based on such chains of conditioned responses. If so, the process of their acquisition would be rather different from the mechanism enabling ants (p. 60) to find their way to and from the nest. If, for example, a mouse is steering a course on the basis of instant information furnished by orienting stimuli which, in the manner discussed on p. 9, release taxes in the form of unconditioned responses, its behavior is very different from that of another mouse which is running along a well-known pathway. The difference between unconditioned and conditioned pathfinding is then dramatically brought to our notice. If such an animal is learning an elevated maze (*Hochlabyrinth*), it proceeds with the utmost deliberation at first, constantly exploring its vicinity with its vibrissae, stopping at literally every step, and even, from time to time, retracing its steps for a short distance. After three or four repetitions, however, the palpitating-palpating procedure is interrupted at certain places by short stretches of smooth, fast running. These stretches, although regularly beginning at the same points of the animal's course, tend to get longer and longer with each repetition. New ones appear; the stretches touch and merge and, when the points of fusion are well smoothed out, the mouse flashes along its path in one single, highly skilled sequence of motor coordinations. Occasionally, definite points of slight hesitation persist stubbornly, much as they do in children reciting poems.

When an animal acquires a path habit under the conditions of its natural habitat, these two processes (first, the slow palpating advance obviously serving the gain of detailed information about the structure of the ground, and second, the smooth run, "skilled" on the basis of information already collected) are distributed, at least at the beginning of the learning process, between the two directions in which the animal traverses the path. Organisms (pomacanthids, blennies, and other highly territorial fishes, lizards, shrews, small rodents, etc.) which "learn by heart" the courses on which they steer behave in the same manner when displaced into an environment altogether unknown to the animal. They all seek cover in the nearest hiding place immediately. Only after having quietened down does the animal start to explore the new environment. It comes out very slowly for a very short distance which at first hardly ever exceeds the length of its own body, all senses alert, and then suddenly whizzes back into cover, quite as if a strong stimulus eliciting escape had impinged. Confidence increases with the mastery of longer and longer stretches of skilled locomotion leading back to safety, until at last all points of the new environment are connected by perfectly smooth path habits with the home base of cover which, in most cases, will be exchanged for a more suitable one in the course of the animal's getting to "know" its new territory. In the interests of survival it is obviously most important that the flight to cover should be one single skilled sequence of movements and not slowed down by the reaction times of several orienting mechanisms. It is of lesser consequence that the way outward should be accomplished as smoothly. Coral fish, when rushing out of their home-base cover in order to expel a trespasser from their territory, occasionally give the impression of using, even on an outward-bound course, a well smoothed-out path habit; but I don't know that I have ever seen a mouse or another rodent rushing along in the typical, skilled manner in any but the homeward direction.

In the first edition of his now classic book, *Die Orientierung der Tiere im Raum* (1919), Kühn described a behavior mechanism which he termed *mnemotaxis,* that is, orientation by memory, a mechanism which he later on regarded more as a theoretical possibility than as a process occurring in reality. He left the word unmentioned in the later editions of his work. The general principle of this orientation mechanism is that of a chain of engrams of extero- and pro-prioceptor stimuli running parallel to and being currently checked by a corresponding sequence of outer stimulations impinging as the organism moves along its path. This is indeed exactly what seems to happen in some animals which keep very strictly to their wonted track. Kühn relinquished the concept of mnemotaxis because many animals can obviously deviate from their track and still not lose their orientedness, and this could not be explained by the principle. Yet in water shrews (*Neomys fodiens L.*), Luther (1936) and I independently observed that the animal has to keep to its track, like a vehicle running on rails, and that it is "derailed" if the slightest detail is changed along the path which the animal has literally "learned by heart." Exactly this must be postulated if Kühn's early concept of mnemotaxis is correct. The path habit of a shrew can be regarded as a chain of locomotor patterns, each conditioned to follow the preceding one and also conditioned to a series of exteroceptor stimuli in agreement with Kühn's concept of "mnemic homophony."

True mnemotaxis, however, seems to be a rare, special case among the mechanisms enabling animals to steer a course by known landmarks. Unlike the shrews, the majority of visually orienting organisms are independent of the animal's position in space as long as pertinent landmarks are in sight. Actually, the latter ability of orientating by means of landmarks implies a computer of a relatively high minimum complexity. The animal's own speed as well as the distance and the angular shift of the visual landmark must necessarily enter into the calculation whenever an organism proves able to steer a straight course by a single landmark, which certainly even insects are perfectly able to do. Tinbergen proposed the term *pharotaxis* for this type of orientation.

Multiple pharotaxis, which enables the animal to ascertain its own position as well as that of each of the landmarks from any possible point within the space known to the animal, is functionally equivalent to spatial "insight" into the structure of the area in question. There are all possible gradations between this way of mastering the problems of space on the one hand, and, on the other hand, the mere "learning by heart" of all possible pathways in a given area. Similar relations may prevail between both of them and true spatial insight. An element of learning and retaining is certainly contained in all the unanalyzed processes which, for lack of a better term, we call insight, as there is necessarily an appreciable time lapse between the reception of stimuli emanating from objects in space and the process of their central representation.

MOTOR LEARNING

In the animal kingdom, receptor learning not only appears at a much lower level of neural organization but generally plays a much more important role than does the learning of new motor patterns. To the best of my knowledge, Storch (1949) was the first to draw attention to this fact. All the specific functions of learning hitherto discussed are based on adaptive modification of processes within the receptor, or afferent side, of neural organization. Possibly the mnemotactically controlled path habits of shrews (p. 69) can be regarded as an exception, because, in their case, the several, primarily independent, elementary locomotor coordinations are welded into one single, coordinated sequence which is so much of a unit that it may justifiably be called one learned movement. It is a question of arbitrarily chosen definition whether or not one decides to do so.

Up to the level of birds and "lower" mammals, all "motor learning" is of this type. Recognizable motor elements, each of them phylogenetically adapted as such, are welded into one "skilled" sequence, and the adaptive value achieved by learning lies in the unhesitatingly smooth way in which the elements follow each other unhampered by interstices and unretarded by reaction times elapsing between them. Even in birds, the complete mastery of the difficult aerial maneuvers which are necessitated by the complicated structure of obstacles in space is often accomplished in this manner. Cave-nesting ducks, for instance, often spend many weeks in learning how to effect a landing in an unfavorably located cave entrance, and the duration of this process is provided for in a simple and effective way in the system of actions of the species in question. "House hunting" begins months before egg laying.

I strongly doubt that the motor coordinations of phylogenetically adapted motor patterns are at all modifiable by learning. I even regard as unproved the belief that an improvement is affected by exercise, comparable to the "breaking in" of an engine. This assumption is as difficult to support as to disprove. If the maturing organism has every opportunity to exercise and a gradual improvement of motor patterns is observed, this might just as well be due to pure maturation. If, on the other hand, the young animal is deprived of that opportunity and the development of certain behavior patterns shows an impairment, the possibility can never be excluded that the latter is brought about by atrophy caused by inactivity.

To what extent a phylogenetically adapted motor pattern as a whole can be brought under the influence of conditioning is a question which has been investigated in a successful experiment by Verplanck (1954). He told me about this experiment some years ago, but surprisingly he never published the results. He tried to reinforce a frequent motor pattern of mallards, the sideways shaking of the bill, by offering a reward immediately after the bird had performed it. The mallards never learned to produce a normal bill shaking "at will," but as soon as they saw the experimenter bending over the pond, they started to perform queer convulsions of the neck muscles, obviously the nearest approach to bill shaking they could achieve by way of learned motor coordinations. From the point of view of their equipment of voluntary movements, these mallards were requested to do something that to them was exactly what Yoga is to man.

Truly "new" learned movements, in the sense of Storch's term *Erwerbsmotorik,* can evidently be developed only by very few animals, and even in these cases, the apparent novelty of these coordinations only derives from the fact that the unlearned elementary motor units of which they are composed are very small indeed. These elements are exactly those which are traditionally called voluntary movements in man. Dissociated, or to be more exact, unassociated muscle contractions are at the beck and call of the pyramidal system and can, therefore, be directly and independently activated by our will. They are the raw material which motor learning strings, like beads on a thread, into a sequence and welds them into one skilled movement. The process by which learning achieves this does not, on principle, need to be different from that which creates a fixed path habit (p. 67). Curiously enough, it seems that the beads which are strung into modifiable sequences by learning are, to a great extent, gained during phylogeny by unstringing phylogenetically adapted fixed sequences of coordination. The study of locomotor coordinations in primates makes this appear highly probable, and this hypothesis offers an unconstrained explanation for the fact that those animals which climb by using prehensile hands have developed a maximum of voluntary movements. As literally every single step and grip of the prehensile limbs must be spatially directed in its smallest detail, it is obvious that the length of fixed motor sequences must be reduced to a minimum and the mechanisms of spatial orientation expanded to a maximum. The "lack of rhythm," which is so conspicuous in exploratory movements of the great apes, is a consequence of this.

Purely voluntary movements, for example, left-handed drawing by a right-handed person or any complicated manipulations performed for the first time, invariably look extremely awkward as long as they are not integrated by learning into a skilled pattern. The voluntary movements are, in this respect, the very opposite of the finished product, although it was the survival value of the latter that caused them to evolve phylogenetically. It is an interesting problem to ascertain the factors which are at work during ontogeny to weld them into the perfection of a skilled pattern.

Probably all the highly developed animals which possess a considerable supply of voluntary movements also possess special built-in mechanisms whose function it is to reinforce the patterns achieving a maximum effect with a minimum expenditure of labor. The existence of such a mechanism can be inferred from its effect; its physiological nature can only be a matter of speculation. It is conceivable that it just correlates proprio- and exteroceptor reports on the labor performed by the muscles with those on the motor effect achieved. This information alone would be sufficient to pick out for reinforcement, among many possible combinations of voluntary movements, those of maximal efficacy and to condemn all less economical ones to extinction. Claiming that self-observation is a legitimate source of knowledge, however, I would suspect that processes closely akin to *Gestalt* perception are also at work, reinforcing smooth, "elegant," efficient, simple solutions of the mechanical problem posed to the organism. Von Holst has shown that the joint effects of relative coordination and the attraction effect (*Magneteffekt*) can achieve this particular task on a very low level of integration. In addition, it can be shown that elementary voluntary movements of man are subject to the very same coordinating influences as the fin movements of a spinal wrasse (*Labrus*). Although in the latter case all of the coordination is dependent on the two factors mentioned, they can play only an auxiliary role in the former, in which higher centers authoritatively decide what motor elements are to be coordinated with what others and in which sequence.

Admittedly as a pure speculation, I would assume that

there are mechanisms in existence which reinforce economical perfection in motor skills independently of the attainment of the ultimate biological goal in whose pursuit the learned movement is developed. This assumption can account for a number of phenomena that are left unexplained by current theories. For one thing, it would give an explanation for the subjective phenomenon to whose importance Bühler was the first to draw attention. What he termed *Funktionslust,* the joy in functioning, is very probably the subjective correlate of what I propose to call the "perfection-reinforcing mechanism" at work. We know that when we skate, ski, or dance, we perform these activities for their own sake, for the enjoyment they give us. If we refuse to accept introspection as a legitimate source of knowledge, we can objectively ascertain that performance of the sports in question by our friends exerts an indubitable reinforcing effect on their behavior just as Harlow and Meyer (1950) have shown that the pure performance of acquired motor skills acts as strong reinforcement on the behavior of monkeys. The existence of motor-perfection–reinforcing mechanisms, like that of many similar receptor organizations, can be more securely inferred from its by-products, which are devoid of survival value than from the "real" function in whose service these mechanisms have evolved.

As human astuteness finds supranormal stimulation for most releasing and/or reinforcing mechanisms (p. 76), it is in no way surprising that it does the same for these mechanisms. Most of these ways happen to be, under the circumstances of civilization, distinctly salutary to our health, which makes it easy to overlook their causal and physiological affinity to vices (p.17). I know one well-known artist and scientist, however, whose addiction to gliding is regarded as a vice by some of his co-workers, because it does occasionally keep him from work. On the other hand, it seems quite possible that, in addition to sports and dancing, all human arts derive at least part of their motivational energy from the very same source.

Proprio- and exteroceptor responses for coordinative and configurational perfection are by no means clearly distinct from each other, and our enjoyment in our own skilled movements is closely akin to the one we derive from watching the skill of others. "The entrancing beauty of everything done superlatively well," as Beebe so very aptly has put it, is inherent not only in the movements performed in the doing, but also in the things done or made by them. It would be easy to lose oneself at this point in philosophical speculation: receptor mechanisms are evolved under the selection pressure of coordinating voluntary movements in the most efficient and perfect way possible. Like all receptor organizations of their kind, they are susceptible to supranormal stimulation and, therefore, keep pushing reinforcement unreservedly in one direction. For once, however, this trend does not end up in misfunction and danger to survival but draws man into the infinite in his appreciation of beauty. One should, of course, have to consider seriously the possibility that our aesthetic judgment of values may perhaps be nothing more than the subjective side of the functioning of the reinforcing mechanisms here under discussion. From the viewpoint of the well-disciplined psycho-physiological parallelists, which we claim to be, the existence of a deterministic and even evolutionistic explanation of our judgment of values would not detract in any way from the reality of the values themselves.

This explanation, however, need not be the only one.

We perceive indubitable beauty and harmony in things which definitely do not owe them to any mechanism promoting motor perfection; phylogenetically adapted motor coordinations, such as the swimming of a shark or the gallop of an antelope, are vying in beauty with the most highly perfected skilled movements. I still believe and claim that I can see something particularly entrancing in a raven's playful soaring antics, in a gibbon's amazing gymnastic stunts, and in the streamlined flourishes of a porpoise, and that I saw it long before I had realized that these movements contained an exceptionally large share of individually acquired skill.

The most convincing argument in favor of my speculative assumption lies in the fact that acquired motor skills of this type, more than any other types of movements, are forever being performed for their own sake in the obvious absence of any other motivating or reinforcing factors. Indeed, the very concept of play is based on this fact to a large extent.

Furthermore, this assumption can give at least a tentative explanation for an otherwise absolutely inexplicable set of phenomena. If a porpoise (*Tursiops*) learns, without intentional training by its human keepers, to throw stones at unpopular persons or to keep an aluminium plate balanced on the tip of its snout while swimming swiftly, the provenience of these highly skilled movements is as great a riddle as is the function of the huge convoluted forebrain of these creatures. The possible explanation which I venture to suggest is that the ancestors of whales were, at the time they began to revert to aquatic life, carnivora with comparatively highly developed brains and were endowed with a fair portion of voluntary movement. When, consequently, a selection pressure was brought to bear on the development of motor patterns that were entirely new to a terrestrial animal, some of it, instead of creating new phylogenetically adapted fixed motor patterns, caused a higher development of voluntary movements which, after all, could be turned to any purpose, including efficient swimming techniques. This would account for the fact that aquatic mammals, for example otters, sea lions, and whales, are able to produce such an amazing multiformity of newly created and elegant, skilled movements in their "play."

The proposed hypothesis that whales have evolved their huge, human-like brain under the selection pressure of no other function than the coordination of swimming may seem a paradox, but the principle voiced by Kipling, "If I have to chop wood with razors, I prefer the best cutlery," is one quite often followed by evolution. In addition, the by-products of voluntary movements which we indubitably observe in Cetacea, and which astonish us so much by their obvious affinity to types of behavior which we consider as characteristic only of our own species, are exactly what one would expect if special mechanisms were at work reinforcing perfection of motor patterns for perfection's sake.

6

Critique of the Earlier Ethologists' Attitude

The contrasting of the "innate" and the "learned" as mutually exclusive concepts is undoubtedly a fallacy, even without considering for the moment the error of applying the first of these two terms to behavior patterns and/or organs (see pp.1, 41). As has been explained in the critique of both behavioristic arguments, it is perfectly possible that a particular motor sequence may owe to phylogenetic processes all the information on environment underlying its adaptedness and yet be almost wholly dependent on individual learning for the "decoding" (p. 20) of this information. This, indeed, is the important truth contained in the second behavioristic argument. The "decoding" of genome-bound information is, in such a case, achieved in two steps: first, by means of morphological ontogeny producing structure; and second, by means of trial-and-error behavior exploiting structure as a teaching apparatus. Processes of this kind are made to appear highly probable by the findings of Prechtl and Knol (1958) on motor patterns and "reflexes" of children born in abnormal positions.

The fallacy of treating the "innate" and the "learned" as mutually exclusive concepts led the old ethologists to make the same errors of which, from page 7 to 21, I have accused behaviorists. Like the behaviorists, the old ethologists took it for granted that the modification of behavior effected by learning invariably caused an increase in survival value. The fact that a phylogenetically evolved neurophysiological mechanism must lie behind this highly differentiated function obviously never occurred to them any more than it did to those of the behavioristic school.

For ethologists as well as for behaviorists, this truly atomistic attitude was a serious obstacle to the understanding of the relations between phylogenetic adaptation and adaptive modification of behavior. It was Lehrman's (1953) critique which, by a somewhat devious route, brought the full realization of these relations to me. In chapter 4 I have discussed the hypothesis that the passive movement imparted to the chick embryo's head by the beating of its heart might take part in teaching the bird to peck. As I have explained, this assumption neglects consideration of the necessity to explain the fact that the motor pattern fits environmental requirements only encountered later in life. In my first counter-critique I said rather satirically that this hypothesis, in order to circumvent the necessity of assuming an innate, that is, phylogenetically adapted motor pattern, unwittingly but unavoidably postulates the necessity of an "innate schoolmarm." In other words, it requires a phylogenetically adapted teaching mechanism. It soon dawned upon me, however, that this thrust against the preformationistic views on learning held by the behavioristic school recoiled with full force on the views held by the older ethologists including myself. I came to realize rather late in life that "learning" was a concept illegitimately used by us as a dump for unanalyzed residue and that, no better than our criticized critics, none of us had ever bothered to ask why learning produced adaptation of behavior. All older ethologists, with the outstanding exception of Craig, had been confining their attention to the innate, while more or less neglecting all problems of learning, without realizing that in doing so they were neglecting one of the most important functions of the majority of phylogenetically adapted behavior mechanisms: the function of teaching!

One of the worst repercussions which this complete neglect of the relations between the "innate schoolmarm" and individual learning had on the real understanding of phylogenetically adapted behavior mechanisms themselves was a one-sided and even physiologically erroneous view of the consummatory act. For a long time it was believed that the gradual rise and final critical drop of excitation typical of all consummatory acts was due to complete exhaustion of action-specific potentiality on the motor side. It has been conclusively demonstrated by experiments of Beach (1942) that in the case of the copulatory activities of the male chimpanzee, it is the effect of reafferent feedback which terminates the consummatory act. This proprio- and exteroceptor feedback is undoubtedly of enormous consequence for the reinforcing function of most or all consummatory stimulus situations. In the case of higher animals, it is hardly an exaggeration to say that the special structure of a typical consummatory act is quite as much the result of the selection pressure exerted by its function of reinforcing appetitive behavior as of the one exerted by its primary function. The lack of appreciation of this fact hindered a true understanding of the processes discussed on p. 64, 65, particularly of the manner in which important information concerning the biologically "right" situation can be contained in the motor pattern itself and can be "decoded" by the animal itself, thus creating a maximum of adaptivity and economy of behavior, as illustrated by the experiments of Eibl-Eibesfeldt (pp. 64, 65). Only Craig gave, at least in his very vivid descriptions of consummatory activities and situations, a hint of the importance of these interactions between phylogenetical adaptation and adaptive modification of behavior (p. 64).

If these admittedly atomistic errors committed by my teacher Heinroth and myself did comparatively little damage to the initial progress of ethology, this is due to the fact that the all-important element of behavior, the fixed motor pattern or *Erbkoordination,* is an extremely invariable skeletal element which influences the system into which it is built far more than it is, in its turn, influenced or changed by the general systemic interaction. The more a particulate element of a system bears the character of such a "relatively independent part" (Lorenz, 1950), the less damage is done by the otherwise illegitimate procedure of isolating it in conceptualization as well as in practical experiments. In any case, by discovering the fixed motor pattern, Whitman and Heinroth created the Archimedean point on which all ethological research is based. Naïvely calling the fixed motor pattern "innate" has done no serious damage to this research in the past or now. On the other hand, the assertion that even the fixed motor pattern can, at least in principle and to an infinitesimal extent, be changed by learning is not based on any observation or experimental result but exclusively on a prejudice and does do real damage, or at least it would do so, if modern ethologists really drew the last bitter consequences from it.

7

The Value and the Limitations of the Deprivation Experiment

All that I have said hitherto in an attempt at a clarification of the concepts of phylogenetically adapted mechanisms, of learning, and of the true relationship between the two is not only the result of an inescapable logical deduction, but it is also, as far as I can see, the only hypothesis that can be brought into agreement with the known facts. It is, furthermore, one that is subject to proof or rejection by the experiment of withholding from the young organism information concerning certain well-defined givens of its natural environment. The rehabilitation of this type of experiment, which for the sake of brevity I propose to call the deprivation experiment, is an important aim of this book. I consider it a serious reproach to the opinions here criticized that they discourage, for reasons gained only deductively from erroneous preconceived ideas, an all-important method of analysis.

The deprivation experiment would be entirely meaningless if learning within the egg or *in utero* could explain the adaptedness of behavior to environmental givens encountered only later in life. Some psychologists tend to assume this, thus unwittingly sharing von Uexküll's postulate of a prestabilized harmony between organism and environment (pp.7-21).

The deprivation experiment would be bereft of any chance to achieve relevant results if one should follow the proposal made by Jensen, and sacrifice, on the altar of the simplest possible experimental and mental operation, the biologist's well-established knowledge that behavior is adapted to environmental givens and that it only can be thus adapted by two well-defined processes, phylogenetic adaptation and adaptive modification (pp. 29–78).

The deprivation experiment would be devoid of value if the modern ethologist's assumption of unlimited and diffuse mutual permeability of phylogenetic adaptation and adaptive modification were correct. Were it so, one could not expect to find a finite number of clear-cut phylogenetically adapted neurosensory mechanisms which, although learning does not in any way "enter into them," nevertheless perform the important function of teaching. This they do in exactly that sense in which I spoke, in erroneous satire, of an "innate schoolmarm." The mistake of overlooking this supremely important function of the "innate" was also made by the old ethologists, who thus missed at least half of what the deprivation experiment can tell us (pp. 79–82).

In other words, the deprivation experiment only makes sense as a tool which can be used to ascertain whence the organism has obtained the information underlying a given adaptedness of behavior. It would immediately disprove the hypothesis of elaborately adapted behavior mechanisms independent of learning if these did not really exist.

The fact that the stickleback "knows" that his rival has a red belly, or that an *Apistogramma* larva "knows" that its mother is striped black and yellow, or that a pup possesses the whole sequence of actions useful in burying a bone in earth without ever having had any previous experience is sufficient to prove that assumption once and for all. We are therefore completely justified in using the deprivation experiment to investigate not only what is not learned, but also the way in which unlearned behavior mechanisms are effective in teaching. In doing so, however, certain technical rules, whose necessity can easily be deduced from what has already been said, must be observed.

THE FIRST RULE OF THE DEPRIVATION EXPERIMENT

The deprivation experiment, taken by itself, can justify direct assertions only about what is not learned. If we withhold from our stickleback all information on the colors of its rival and if it nonetheless does respond to that color pattern, we then know without any further experiment that information concerning the rival's colors has been relayed by the genome. If, on the other hand, our subject would have failed to show a specific response to its rival's nuptial coloration, we should not be justified in asserting that this response is normally dependent on learning. There would still be the alternative explanation that, while trying to withhold from the animal information only, we have either inadvertently withheld "building stones" indispensable to the full realization of the blueprint contained in the genome or else that we are withholding in the experimental setup a stimulus situation necessary to release the behavior we are undertaking to investigate.

The plain statement that, for the reasons given, the

deprivation experiment permits immediate assertions only concerning what is not learned and not what is learned has been branded by some behavioristic psychologists as "a highly protective theory" insiduously devised by us to shield the concept of the innate against the inroads of analysis proving the non-existence of neural mechanisms into which learning does not enter. It has been argued that the above-stated limitation of the deprivation experiment precludes all possibility of ever proving definitely that any given adaptedness of behavior is dependent on learning, because there always remains a chance that, given other environmental conditions, the behavior element under investigation might prove to be phylogenetically adapted. Thus, the argument continues, the "hypothesis" that a given adaptedness of behavior is phylogenetically evolved is irrefutable by experiments and, hence, not analytically valid.

As this highly sophisticated argument has made a surprisingly deep impression, I think its explicit refutation is required. For one thing, the argument wrongly presupposes that one single experimental setup must necessarily be able, quite by itself, to distinguish between phylogenetic adaptation and adaptive modification, and furthermore, that no other method exists to do this apart from the deprivation experiment. Just as in chemical analytical procedure, one test is only expected to prove the presence or absence of one particular substance, the deprivation experiment cannot be expected to achieve more than to ascertain, under favorable circumstances, the presence of phylogenetic adaptation. Even if an experiment can only give a partial answer to a yes-or-no question, this makes it analytically valid. It is indeed a widespread and dangerous atomistic error that all analytic approaches to living systems can be formulated in such a way as to receive either "Yes" or "No" for an answer. It is fully legitimate, even in the so-called exact natural sciences, to state quite simply, "Under such-and-such circumstances this or that happens. If I let go of the stone, it will fall down to earth with the acceleration *g*." or "If I provide the rat with nesting material, it will learn to perform carrying, heaping up, and upholstering movements in the order named. If I do not provide it with material it will not develop this order." All these statements are perfectly legitimate in natural science, irrespective of the faint possibility that the stone will do something quite different or of the equally remote chance that the rat, under the influence of some miraculous drug, might hit on this well-adapted sequence of behavior patterns.

Moreover, though the deprivation experiment does not tell us what is learned, it does tell us what some of the most important teaching mechanisms are. If we see a pup perform its bone-burying activities in the drawing room, we do not need much imagination to take this chance observation as a hint to tell us where to look for the stimulus situations which might act as reinforcements that are teaching the animal to perform the activity. We can at once make a good guess at the situation in which it attains its survival value. If we see an inexperienced rat performing the heaping-up or upholstering motor patterns in the air (Eibl-Eibesfeldt's experiments), we get a strong indication of where to look for the reinforcing consummatory situations which teach the animal to integrate its several fixed motor patterns into the "correct" sequence—in other words, the one whose survival value exerted the selection pressure which, in turn, caused the whole phylogenetically adapted mechanisms to evolve.

Forty-five years have now passed since Craig in his classic paper (1918) clearly emphasized that one of the important functions of all consummatory acts lies in reinforcing whatever behavior preceded it. Few psychologists of the behavioristic school have as yet drawn the obvious conclusion that, in order not to overlook unsuspected reinforcements of extreme effectiveness, one must know all consummatory acts contained in the investigated subject's system of action. Few ethologists, until quite recently, have realized that one must know quite a lot about the learning processes of a species in order to appreciate the teaching functions of consummatory acts and the selection pressure which these functions exert on their structure.

The teaching function of the consummatory act is based partly on phylogenetically acquired information contained in the motor pattern itself and partly on that contained in the receptor organization of the releasing mechanism but mainly on the interaction of both, that is to say, on the reafference which the organism produces for itself by performing the consummatory act in the adequate consummatory situation. This specific interlocking of motor and receptor patterns achieves not only a high specificity and selectivity but, by a mechanism of positive feedback, a particular intensity of the reinforcing effect which we recognize as the most important function, to whose selection pressure the mechanism under discussion owes its characteristic organization.

The motor pattern performed in its complete and immediately recognizable sequence by the deprived organism tells the observer at once and automatically how much of it is not learned. The function of a receptor organization, such as the stickleback's responding to "red below," only tells us that there is something that is not learned, something sufficiently characteristic of the natural object of the response. The definite response of our subject to the configurational sign stimulus of red underneath gives us no indication, however, as to whether the organism possesses any additional phylogenetically acquired information concerning the natural object of its rival-fighting response. The green eye, the blue back, the threat postures, and other properties of the rival might be sign stimuli as well, as indeed they are; but each of them has to be investigated separately by a technique which excludes the danger of mistaking for innate information all that which is really the result of a very quick learning process. This method was mentioned in chapter 5 and will be dealt with in the third rule of the deprivation experiment.

Owing to the fact that it can only tell us directly what is not learned, the deprivation experiment affords the most far-reaching shortcuts to the analysis of behavior in those cases in which it proves very complicated and elaborately adapted systems to be independent of specific, individually acquired information. A good example is furnished by the orientation mechanisms studied by Hoffmann (1952) in the starling (*Sturnus vulgaris L.*). A starling which has never seen the sun or its movement across the sky possesses not only perfect information on how the sun moves when viewed from the northern hemisphere but also possesses a chronometer and a computer enabling it to get its bearings from the position of the sun at all hours of the day. Phylogenetically adapted mechanisms of this kind are particularly susceptible to damage wrought by insufficient general health of the subject, and tribute is due equally to the astuteness of the scientist in arranging his experiments and to his proficiency in the high art of rearing and keeping birds.

THE SECOND RULE OF THE DEPRIVATION EXPERIMENT

A very important rule which has to be considered in every deprivation experiment follows logically from what has been said in criticism of some modern ethologists and in the last chapter. The rule is simply that the investigator must possess a very thorough knowledge of his subject's whole system of actions, or "ethogram," as well as of the pathological symptoms which are "produced by other operations (other than selecting or training)" and which, as I have shown (pp. 29–78), are likely to disintegrate that system. For the purpose of expounding the methodological rules governing the deprivation experiments, I propose to substitute for the concept of these "operations" the simpler one of "bad rearing."

Of course, withholding specific information is also one of the operations which disintegrate an animal's behavior system. The great horned owls (*Bubo bubo L.*), the Hungarian partridges (*Perdix perdix L.*), and many other birds which the Heinroths reared for their classic work (1924–28) proved unable to breed because all their sexual responses had been irreversibly fixated on human beings by the "operation" of letting humans, instead of conspecifics, do the rearing. Naturally this disturbance of adapted behavior is just as much a pathological condition as any wrought by bad rearing. But the point is that we can differentiate the effects of both by their respective symptoms! Any tolerably good medical man would, even without knowledge of the pre-history of the subject, deduce that very different subsystems of the central nervous system have been affected in each case.

In all cases in which information necessary for the adaptive modification of a certain behavior unit is withheld or misinformation substituted for it, only that one unit is affected. Large subsystems, even those which take part in the same survival function, remain entirely undisturbed. That the investigator must literally know by heart the ethogram of his subject is a postulate which derives from the necessity of recognizing such "chunks" of undisturbed behavior patterns on sight. On the other hand, our conviction that phylogenetic adaptation and adaptive modification of behavior do not diffusely mix derives from the fact that in deprivation experiments in which bad rearing is avoided we always find such undisturbed chunks. The famous puppy performs its burying activities in a perfectly coordinated movement, indistinguishable from that used by the experienced control; the same is true of the rat's single motor patterns of nest building. The mandarin duck imprinted on mallards will, in spite of the pathologically induced choice of object, perform the whole complicated system of courtship activities in a manner identical with that of normal conspecifics.

On the other hand, in practically all cases in which operations, other than those preventing the input of specific information, disrupt the system of actions (in other words, in those cases in which non-specific damage is wrought by bad rearing), we find more than one system of behavioral mechanisms disturbed. If we rear any visually oriented organism, for example, a stickleback, in total darkness, we might find that this animal fails to respond to the property "red below" in a dummy or a conspecific. But we are sure to find that other equally selective visual responses are equally impaired and, for that matter, a number of other seemingly unconnected behavior mechanisms as well.

Although non-specific with respect to the systems

affected by it, bad rearing produces some very typical and easily recognizable symptoms in certain types of phylogenetically adapted processes in behavior. At least three of these symptoms must be known to the investigator performing deprivation experiments:

1. Fixed motor patterns depending on endogenous production of excitation, particularly those which occur only rarely in the animal's life, such as fighting or activities belonging to the reproductive cycle, are very sensitive to bad rearing and bad keeping and respond to both by a quantitative decrease. This is the reason why we so very often see, in captive and hand-reared animals, only rudiments of specific activities. For example, nest building occurs in captivity much more often in the form of incomplete beginnings than in a fully adapted, completed sequence. Failure to breed a species in captivity is in the majority of cases caused by this phenomenon. The less pronounced a quantitative decrease in the performance of fixed motor patterns, all the more dangerous does it become as a source of error. In particular, it makes it very dangerous to ever assert that a certain species is lacking a certain motor pattern. Very slight damage caused by bad rearing may not be apparent in any other behavioral deficiencies. Yet it may cause some motor patterns which possess a high threshold of specific excitation to disappear.

This is one of the important exceptions to the rule that bad rearing always causes several systems to disintegrate. In my paper on the motor patterns of social courtship in dabbling ducks, *Anatini,* (1942), I described a number of cases in which a pattern was lacking in a species in which later studies by von de Wall (1963) showed it to be present, although very rare indeed. The discrepancy was indubitably caused by the rather primitive conditions under which the subjects of my former observations were reared and kept in Altenberg. They simply failed to reach the thresholds of action-specific excitation which are necessary to set off the patterns under discussion.

2. Another very typical disturbance caused by bad rearing consists of the loss of normal selectivity by phylogenetically adapted releasing mechanisms. In certain cases, this phenomenon can counteract the motor weakness of an activity by making its elicitation easier, but the indiscriminate way in which the activity is then performed in obviously "wrong" situations helps to diagnose it.

The loss of selectivity caused by bad rearing in some releasing mechanisms can, however, be very misleading indeed. A brood of red-backed shrikes (*Lanius collurio*), which I reared many years ago, began to perform wiping movements along their perches with pieces of meat or other chunky food tightly held in their bills soon after fledging. I correctly suspected that these were the beginnings of the motor patterns with which shrikes impale larger prey on the thorns of certain plants in order to store food. This was confirmed when I offered my birds artificial thorns in the form of nails driven through the perches. These "thorns" protruded vertically upward about 2 cm. At first, the shrikes paid no attention to these new objects. After a bird had once happened to hit a nail in a wiping movement and catch its prey on the protrusion, it rapidly increased the intensity of this movement, thus succeeding in really impaling the prey on the thorn. Afterward, the bird very quickly learned to direct its wiping movement at a nail. When Kramer repeated the same experiment with red-backed shrikes which he had reared from his research on bird navigation, he found full confirmation of what I have just described at first; but when he improved his feeding technique by adding live silkworms, bred for that purpose, to the rearing diet of his next batch of shrikes, these birds did not need any trial-and-error behavior to condition them to the thorn. They immediately went for the thorn when it was first offered. When offered a rubber dummy of equal shape, they persisted in directing their impaling movement at that unsuitable object. Lack of success was unable to effect a negative conditioning (Kramer and von St. Paul, unpublished).

3. A highly typical symptom of "bad rearing" is the disintegration of social inhibitions. It is known all too well by all animal breeders and zoo men that carnivorous and omnivorous animals, when breeding in captivity, are prone to eat their own young. Saying that they do so "as often as not" is hardly an overstatement. The opinion, shared by many experts, that very tame females are particularly prone to this cannibalism, is true in itself. It is not, however, the tameness of the individual that causes the disturbance, but its antecedents. Being hand-reared, or at least caught very young, may very easily cause the animal to become familiar with man and may also cause it to sustain slight bodily damage.

THE THIRD RULE OF THE DEPRIVATION EXPERIMENT

Another rule concerns an important difference in the methods of experimental approach to phylogenetically adapted motor patterns and to releasing mechanisms. A motor pattern can fit an object which the animal has never as yet individually encountered in so many mutually independent points that a correspondingly great number of phylogenetically acquired pieces of information becomes directly apparent in one single experiment. The puppy which tries to bury a bone on the parqueted floor has perfect innate information that it is a good thing when food is left over to dig in a secluded corner, deposit the leftover, and finally to push back, by convergent shoving movements of the nose, whatever material has been dug up. If, on the other hand, a releasing mechanism contains innate information concerning several characteristics of the biologically adequate situation, one deprivation experiment can ascertain, on principle, only one point of innate adaptedness. If we present to our stickleback, which has never in its life encountered a male rival, a dummy which is red below, has green eyes, a blue back, and which stands head downward while making jerking movements, and if our subject responds at once by performing the specific motor patterns of rival fighting, we can only say that it had innate information about at least one of those characteristic properties of its rival which are presented in the model. For this rather hazy result we have sacrificed our carefully reared subject. Henceforward it has become useless for any further experimentation to find out on which of the properties of the first dummy it has innate information. We know from experience how rapidly the innate information contained in a releasing mechanism effects conditioning. If an inexperienced animal is exposed to a complex situation containing, among others, stimuli on which the subject possesses innate information, the latter may effect a conditioning to all the stimuli presented simultaneously (Schleidt, 1962).

Innate information on the receptor side—not on the side of motor patterns—is always couched in extremely simple terms. It is a fundamental error to think that selectivity of a response indicates the presence of innate information. The exact opposite is true. It is quite a reliable rule of thumb, which I stated as early as 1935, that if the animal is taken in by simple dummies one can assume that the underlying information is innate, while, conversely, high selectivity is practically always an indication that learning has played a part. This rule can, of course, only be used as a very general pointer indicating where to look for innate and where to look for acquired information. It does not, by any means, make the deprivation experiment superfluous.

The misleading impression that innate information is more specific than it really is often arises when one neglects the possibility of extremely quick "flashlike" conditioning, which may be even quicker than in Schleidt's example of the turkey hen (p. 36). The ever-present possibility of such rapid conditioning casts a serious doubt on many conclusions, drawn from deprivation experiments, in which this source of error is not considered. Rereading the experimentation hitherto performed on the stickleback, I find that it even seems uncertain that the classical example for innate information, "red beneath," is really innate. Nobody seems to have presented a male stickleback which was reared in isolation with a dummy that is completely red or red above in the first trial. If, on this first occasion of presentation, the dummy was red below, it is quite possible that the innate information is confined to "red" alone and that the configurational property "below" is learned in a flash. The same applies to the results of Beach and Jaynes (1956), who found that the speed with which mother rats retrieve their young is dependent on a combination of optical, olfactory, and tactile properties of the object, and who concluded that in mammals, unlike what is known of fishes and birds, behavior patterns may be released and directed by complicated stimulus situations in which several sense organs take an equal part. Schleidt (1962) points out that the experiments were performed as late as twenty-four hours after parturition, a period amply sufficient to supplement innate information by learning. The authors did not know then of the results of Zippelius and Schleidt (1956), who discovered that the young of many small rodents utter supersonic call notes which immediately induce the mother to search for and retrieve the baby. On the basis of what is known today, it appears quite possible that the response to this supersonic call contains the only innate information which the mother possesses about her progeny, and that all the selectivity, found by Beach and Jaynes, which necessitated other stimuli to release the retrieving response, is the result of learning. It might also be that the response to the smell is innate and that to the call note conditioned. The inexperienced jackdaw (*Coloeus monedula*) reacts innately to the parent's warning call and is quickly conditioned by it to recognize dangerous predators. Conversely, in the young curlew (*Numenius arquata*) the response to flying predators is based on innate information (von Frisch, 1956) and conditions the bird to respond to the parent's warning call.

This third rule, which must be obeyed in all deprivation experiments which are concerned with the innate information underlying releasing mechanisms, can be summarized in the words of Schleidt, "We are forced to use each animal in one experiment only or else to plot, for each subject, a sequence of experiments, in each of which the possible acquisition of information is controlled in such a way that it can be taken into consideration in the succeeding ones [free translation]."

THE FOURTH RULE OF THE DEPRIVATION EXPERIMENT

The next rule is so easy to understand that it is surprising that it has been disregarded by notable experimenters. The experimental setup which is calculated to withhold certain kinds of information from the subject must be carefully devised as not to withhold the stimulus situation necessary to elicit the behavior pattern under investigation at the same time. Otherwise, one can come in danger of mistaking defects of behavior which are exclusively due to the lack of an adequate stimulus situation for the consequence of withholding specific information.

An example for this is furnished by the experiments of Riess (1954), who reared rats in such a manner as to deprive them of all experience in handling solid objects. When afterward, in a strictly standardized setup, the rats failed to build nests, Riess concluded that nest building is dependent on previous experience in handling solid materials. What he failed to realize was that his test situation was such that no rat, experienced or not, would have started nest building in it within the allotted time.

This type of error is avoided by a simple cross-check: the subject of the deprivation experiment is liberated in an environment closely corresponding to the wild habitat of the species and a normally, experienced control is tested in the experimental setup. When Eibl-Eibesfeldt (1955, 1963) later repeated Riess's experiments with such dramatically different results (p. 64), I had the greatest difficulty in persuading him to put an experienced rat in a test situation modeled as closely as possible on that of Riess. Finally Eibl-Eibesfeldt, then a very young man, did so, self-consciously shrugging his shoulders. He considered my request nonsense because he knew that any rat, even a pregnant one or one with a new litter, will spend at least a day in exploratory behavior and attempts to get out when put into a wooden box which is new to it and devoid of any cover, and will certainly not start building during that period.

THE FIFTH RULE OF THE DEPRIVATION EXPERIMENT

The last rule I should never have dreamed of formulating here, if gross infractions had not been committed by highly qualified investigators. It says that agreeing results can only be expected if subjects of tolerably similar genetical constitution are used. Geneticists have correctly emphasized that wild animals cannot, strictly speaking, be propagated in captivity, because captivity changes all hitherto effective selective factors in so profound a manner that serious changes must be expected in the genome of the stock after only a few generations. These changes can cause, in phylogenetically adapted behavior mechanisms, defects strongly resembling those brought about by the slightly pathological effects of bad rearing (p. 90). Some cichlid fishes have been bred by commercial breeders in a manner so artificial that the activities of parental care which these fish normally perform became superfluous. In two species, *Pterophyllum eimeckei* and *Apistogramma ramirezi,* this has caused the partial disappearance of the inhibitions normally preventing the parents from devouring their own eggs and young. Today, it is very difficult to find a normally behaving pair among commercial stock of these fish. In the Texas cichlid (*Herichthys cyanoguttatus*), Leyhausen and I observed, shortly before the war, a very differentiated warning behavior of the female and a corresponding response of the young. In the very small and highly inbred stock of this species which survived the last world war in Germany, both the warning movement and the response of the small fry are entirely absent.

If defects of this sort develop in wild animals which are bred in captivity for only a few years, correspondingly greater ones are to be expected in domesticated species which have been bred for many centuries. In 1937, Tinbergen investigated the phylogenetically adapted response to flying predators in turkeys, pheasants, and greylag geese. In 1955 Hirsch, Lindley, and Tolman (1955) wrote, "The Tinbergen hypothesis that . . . was tested on the white leghorn chicken and found untenable under strict laboratory conditions." This statement is exactly as meaningful as if somebody were to have demonstrated the presence of melanins in the hair of wild common hamsters and if somebody else were to have written: "The Somebody theory that there are melanins in the fur of wild hamsters was tested on white laboratory rats and found untenable under strict laboratory conditions."

8

Summary

The primary aim of this book is to prevent the discrediting of a concept which, though it has been used occasionally in an imprecise manner, is indispensable to an ethological approach. This concept is that of the innate.

The second aim is to prove the validity of an all-important line of research which would be discouraged and even discontinued if this concept were abandoned.

In pursuit of this dual purpose, three widely held attitudes toward the conception of the innate are discussed.

THE ATTITUDE OF AMERICAN PSYCHOLOGISTS

Many American psychologists negate the validity of the concept of the innate, mainly on the basis of the following two arguments:

1. The dichotomy of behavior into innate and learned arises only from begging the question, as one can only be defined by the exclusion of the other (Hebb, 1953).

2. No experimentation can ever ascertain how much of an animal's behavior is innate, as the possibility of its learning within the egg or *in utero* can never be excluded (Lehrman, 1953).

THE ATTITUDE OF ENGLISH-SPEAKING ETHOLOGISTS

Many modern ethologists, particularly those publishing in English, contend that the term innate is not only useless, but heuristically harmful. They assume that phylogenetic adaptation and adaptive modification can be added to and mixed with each other in any behavior mechanism, however minute and elementary. For this reason, they regard it as hopeless and even dangerous to try separating, in experiment or thought, innate and learned elements of behavior. Even if, for example, a stickleback lacking all previous experience with a rival fights a model which is red underneath at first sight, this behavior cannot be called innate because some of its components and prerequisites, such as swimming movements or point discrimination on the retina, may have undergone adaptive modification during ontogeny.

THE ATTITUDE OF THE EARLIER ETHOLOGISTS

Most older ethologists believed—and still do—that there are, in the machinery of behavior, quite considerable self-contained units into which learning does not enter, and which, in the hierarchical organization of appetitive behavior, are intercalated with links that are adaptively modifiable by learning. Although they do not define innate behavior by its not being learned but by its being phylogenetically adapted, these older authors at one time considered the concept of the innate and the learned as being mutually exclusive.

CRITIQUE OF DICHOTOMY BY EXCLUSION

Neither the concept of the innate nor that of the learned is defined by the exclusion of the other. Both are defined by the provenience of the information which is the prerequisite of behavior being adapted to environment. There are but two ways in which this information can be fed into the organic system.

The first is the interaction of the species with its environment during evolution. By mutation and selection, a method analogous to learning by trial and success (not by error!) and also analogous to induction devoid of any deductive procedure, the species gathers information and stores it, coded in the form in chain molecules, in its genome.

The second is the interaction with the environment in which the individual acquires information. There are two distinctive ways in which it does so. One is its instantaneous response to impinging stimuli which orient it in space and time without, however, changing anything in the machinery of its behavior. The other is the adaptive modification of this machinery. This process also implies the storing of information, probably in the central nervous system, although we know nothing about the physiological mechanism of the "engram." For the purposes of this paper, adaptive modification of behavior is provisionally equated to learning.

The chances of random modification being adaptive are not greater than those of mutation being so; the probability is assessed as 10^{-8} by geneticists. Any modifiability which regularly proves adaptive, as learning indubitably does, presupposes a programming based on phylogenetically acquired information. To deny this necessitates the assumption of a prestabilized harmony between organism and environment.

On the other hand, there is no evidence and no logical reason for assuming that all phylogenetically adapted machinery of behavior must unconditionally be susceptible to adaptive modification. Quite the contrary, there are strong arguments in favor of the assumption that certain mechanisms of behavior, for the simple reason that they do contain the phylogentetic programming of learning processes, must themselves be refractory to any modificatory change.

No biologist treats the concepts of phylogenetically adapted and adaptively modified organs or behavior patterns as being mutually exclusive. This dichotomy would indeed be fallacious but in a sense directly opposed to the one implied by Hebb: phylogenetic adaptation must unconditionally be contained in every learning process, while the hypothesis that learning must enter into all phylogenetically adapted behavior is entirely unfounded.

CRITIQUE OF THE ASSUMPTION OF EARLY LEARNING

Although learning can undoubtedly occur while the embryo is still within the egg or *in utero,* it can gather information only on that which is accessible to the growing organism in its present environment. It cannot acquire information on environmental givens which are only encountered in later life. To assume for instance, that the chicken can learn to peck by having its head passively moved up and down by the heartbeat before hatching (Kuo, quoted by Lehrman, 1953) postulates either the existence of a very special phylogenetically programmed teaching mechanism or, again, the assumption of prestabilized harmony.

It is not to be denied, however, that the embryo can indeed learn important elements of behavior before being born or hatched. Information contained in the genome can be doubly decoded, first by morphogeny and subsequently by trial-and-error learning while using morphological structure (Prechtl and Knol, 1958).

CRITIQUE OF THE ENGLISH-SPEAKING ETHOLOGISTS' THEORY

In theory it is correct that practically all environmental prerequisites of phenogeny may cause by their variation slight differences in the ultimate behavior of the organism. It is also true that, among the changes thus effected, there is no sharp borderline between the pathological, the "normal," and that which may be regarded as adaptive modification. Like both the arguments of the American psychologists, however, this theory does not consider the fact that every case of adaptive modifiability presupposes its own, phylogenetically adapted mechanism. As the theory assumes an unlimited number of adaptively modifiable elementary processes, one would have to assume an equal number of such mechanisms. This is obviously contradicted by observational and experimental facts. Also, learning processes are not as easily overlooked in the study of the ontogeny of behavior as the theory implies.

Furthermore, even if an almost unlimited number of adaptive modifiabilities of behavior were to exist, it would still not be necessary to investigate all the effects of all the many factors causing differences in behavior in order to ascertain the innate information on certain givens in the environment of the species. The deprivation experiment need not be concerned with all the prerequisites of normal phenogeny as long as they do not contain the information whose provenience is being investigated. Whatever wonders epigenetical phenogeny may perform, for instance, in the ontogeny of a stickleback, it cannot possibly extract from the factors indispensable for healthy growth (light, oxygen, sufficient food etc.), the information that the rival who must be fought is red on the underside.

CRITIQUE OF THE EARLIER ETHOLOGISTS' THEORY

The only serious fallacy in the older ethologists' attitude is, curiously enough, identical with the one already criticized in both the arguments proffered by American psychologists. They, too, failed to realize that the survival function of learning presupposes phylogenetical programming. In consequence of this oversight, they did indeed, as Hebb pointed out, form mutually exclusive concepts of the innate and the learned. These concepts were not, however, not even at first, defined by mutual exclusion but by the differences between phylogenetic adaptation and adaptive modification. With the notable exception of Craig, the older ethologists did not grasp the teaching function of phylogenetically adapted behavior; hence, they also neglected the selection pressure exerted on the consummatory act by its important function of reinforcing antecedent behavior. If their attempt to study innate behavior without paying any attention to learning merits the reproach of atomism, they did, at least, attribute sufficient importance to the distinction of the innate and the learned.

No insight into the physiological causation of behavior can be gained without knowledge of the source of the information contained in any of its adaptations to environment. Neither the close cooperation between phylogenetic adaptation and adaptive modification of behavior nor the close analogy between the "methods" by which both processes extract information from the environment must ever make us forget the basic difference in their physiological causation. To deny the validity of the distinction of the corresponding concepts would be just as vapid in the field of behavior study as in any other branch of biological research. Nobody ever dreamed of abandoning the concepts of phylogenetic adaptation and adaptive modification in morphogeny, in spite of the fact that both processes are usually superimposed, one upon the other, in one phenomenon, and that, in the case of the phenocopy, the effect of one is indistinguishable from that of the other.

III. THE DEPRIVATION EXPERIMENT

If the information which is clearly contained in the behavioral adaptation to an environmental given is made inaccessible to the individual's experience and if, under these circumstances, the adaptedness in question remains unimpaired, we can assert that the information is contained in the genome. The conclusion to be drawn from that which we call the deprivation experiment is as simple as that! In performing experiments of this type, however, we must keep the following rules in mind:

1. The deprivation experiment can tell us, directly, only of the existence of innate information. Negative results, that is, defects of adaptation as consequences of the experiment, can have causes other than the withholding of information. The participation of learning processes cannot, therefore, be deduced immediately. The nature of the phylogenetically adapted behavior patterns which appear undamaged by the experiment, particularly disjointed consummatory acts, give valuable indications as to where, in

the species' system of actions, learning processes are programmed, which are the reinforcements effecting them, and how the unifying function of learning can be proved experimentally.

2. In order to recognize, for what they are, the disjointed fragments of behavior chains which regularly appear in deprivation experiments and in order to know where they fit into the normal program of the species' behavior, the experimenter must be thoroughly familiar with the system of actions of his subjects. He must also have a good "clinical eye" and must know the symptoms of bad rearing inside out. If he fulfills these requirements, he will hardly ever confound the diffuse damage wrought on many sub-systems of an animal's behavior by bad rearing with the circumscript defects caused by withholding individual information.

3. The procedure of the deprivation experiment must be an entirely different one according to whether a motor pattern or a receptor organization is investigated. In the case of the motor pattern, one experiment may enlighten us on very many single points of innate information on environment. If, on the other hand, a releasing mechanism is being examined, it is necessary to perform as many experiments, each with its own subject, as there are givens in the environment on which the species possesses innate information. Otherwise, there is the danger of mistaking that which has really been learned in one experiment for innate information in a subsequent one.

4. Care must be taken that the experimental setup devised to withhold information does not simultaneously withhold stimulus situations necessary to elicit the behavior which is to be investigated. The simplest way to avoid this danger is to test a normal control under the experimen-

tal conditions and, conversely, the subject of the deprivation experiment under circumstances corresponding closely to those of the species' natural environment.

5. Agreeing results in investigating innate information can only be expected when objects of identical or closely similar genetic constitution are used. One cannot test results obtained in turkeys by using white leghorn chickens.

References

Aschoff, J. 1962. Spontane lokomotorische Aktivität. *Handbuch der Zoologie* 10 (11). Berlin: De Gruyter.

Braemer, W. and Schwassmann, H. Der Sonnenkompass bei sonnenlos aufgezogenen Fischen. In preparation.

Beach, F. H. 1942. Analysis of Factors Involved in the Arousal, Maintenance, and Manifestation of Sexual Excitement in Male Animals. *Psychosomatic Med.* 4:173.

Beach, F. H. and Jaynes, J. 1956. Studies of Maternal Retrieving in Rats. III. Sensory Cues Involved in the Lactating Female's Response to Her Young. *Behaviour* 10:104–25.

Brunswik, E. 1957. Scope and Aspects of the Cognitive Problem. In J. S. Bruner *et al., Contemporary Approaches to Cognition.* Cambridge: Harvard University Press.

Bühler, K. 1922. *Die geistige Entwicklung des Kindes.* Jena: Gustav Fischer.

Campbell, D. T. 1958. *Methodological Suggestions from a Comparative Psychology of Knowledge Processes.* Oslo University Press, Inquiry.

Craig, W. 1918. Appetites and Aversions as Constituents of Instincts. *Biol. Bull.* 34:91–107.

Darwin, C. 1859. *Origin of Species.* London: John Murray.

Eibl-Eibesfeldt, I. 1951. Nahrungserwerb und Beuteschema der Erdkröte (*Bufo bufo L.*). *Behaviour* 4:1–35.

Eibl-Eibesfeldt, I. 1955. Nestbauverhalten der Wanderratte. *Naturwissenschaften* 42:633–34.

Eibl-Eibesfeldt, I. 1955. Zur Biologie des Iltis (*Putorius putorius L.*). *Verhandl. Deut. Zool. Ges. Erlangen* 304–23.

Eibl-Eibesfeldt, I. 1956. Über die Ontogenetische Entwicklung der Technik des Nüsseöffnens vom Eichhörnchen (*Sciurus vulgaris*). *Z. Säugetierkunde* 21:132–34.

Eibl-Eibesfeldt, I. 1961. The Interactions of Unlearned Behaviour Patterns and Learning in Mammals. *Symposium on Brain Mechanisms and Learning.* CIOMS Montevideo. Oxford: Blackwell, pp. 53–73.

Eibl-Eibesfeldt, I. 1963. Angeborenes und Erworbenes im Verhalten einiger Säuger. *Z. Tierpsychol.* 20:705–54.

Frisch, K., von. 1914. Der Farbensinn und Formensinn der Biene. *Zool. Jahrb.* 35:1–188.

Frisch, O., von. 1956. Zur Brutbiologie und Jugendentwicklung des Brachvogels. *Z. Tierpsychol.* 13:50–81.

Harlow, H. F. and Meyer, D. R. 1950. Learning Motivated by a Manipulation Drive. *J. Exptl. Psychol.* 40:228–34.

Hassenstein, B. 1954. Abbildende Begriffe. *Verhandl Deut. Zool. Ges. Tübingen,* 197–202.

Hebb, D. O. 1953. Heredity and Environment in Mammalian Behavior. *Brit. J. Animal Behaviour.* 1:43–47.

Hertz, M. 1937. Beitrag zum Farbensinn und Formensinn der Biene. *Z. vgl. Physiol.* 24:413–21.

Hess, E. H. 1956. Space Perception in the Chick. *Sci. Am.* 195 (1):71–80.

Heinroth, O. 1910. Beiträge zur Biologie, namentlich Ethologie und Psychologie der Anatiden. *Verhandl. Ver. Intern. Ornithol. Kong. Berlin,* 589–702.

Heinroth, O. and Heinroth, M. 1924–28. *Die Vögel Mitteleuropas.* Berlin. Bermühler Verlag.

Hinde, R. A. 1960. Factors Governing the Changes in Strength of a Partially Inborn Response, as Shown by the Mobbing Behaviour of the Chaffinch (*Fringilla coelebs*). *Proc. Roy. Soc. B.* 753, 398–420.

Hoffmann, K. 1952. Die Einrechnung der Sonnenwanderung bei der Richtungsweisung des sonnenlos aufgezogenen Stares. *Naturwissenschaften.* 40:148.

Hoffmann, K. 1954. Versuche zu der im Richtungsfinden der Vögel enthaltenen Zeitschätzung. *Z. Tierpsychol.* 11:453–75.

von Holst, E. 1935. Über den Prozess der zentralnervösen Koordination. *Pflüg. Arch.* 236:149–58.

von Holst, E. 1955. Regelvorgänge in der optischen Wahrnehmung. *Rept. 5th Conf. Soc. Biol. Rhythm., Stockholm.*

von Holst, E. 1957. Aktive Leistung der menschlichen Gesichtswahrnehmung. *Studium Generale* 4:231–43.

Hull, C. L. 1943. *Principles of Behavior.* New York: Appleton-Century.

Jander, R. 1957. Die optische Richtungsorientierung der roten Waldameisen (*Formica rufa L.*). *Z. vgl. Physiol.* 40:162–238.

Jensen, D. D. 1961. Operationism and the Question "Is This Behavior Learned or Innate?" *Behaviour* 17:1–8.

Kühn, A. 1919. *Die Orientierung der Tiere im Raum,* Jena: Gustav Fischer.

Kühme, W. 1962. Das Schwarmverhalten elterngeführter Jungcichliden (*Pisces*). *Z. Tierpsychol.* 19:513–38.

Kuenzer, E. and Kuenzer, P. 1962. Untersuchungen zur Brutpflege der Zwergcichliden *Apistogramma reitzigi* und *A. borrellii. Z. Tierpsychol. 19:*56–83.

Kuo, Z. Y. 1932. Ontogeny of Embryonic Behavior in Aves. *Intern. J. Exptl. Zool.* 61:395–430, 453–89.

Lehrmann, D. S. 1953. A Critique of Konrad Lorenz' Theory of Instinctive Behavior. *Quart. Rev. Biol.* 28:337–63.

Lorenz, K. 1941. Vergleichende Bewegungsstudien an Anatiden. *J. Ornithol.* 89:194–293.

Lorenz, K. 1950. Ganzheit und Teil in der tierischen und menschlichen Gemeinschaft. *Studium Generale* 9:455–99.

Luther, W. 1936. Beobachtungen an einer gefangenen Wasserspitzmaus (*Neomys fodiens Schreb.*). *Zool. Garten* 8:303.

Massermann, J. H. 1943. *Behavior and Neurosis.* Chicago: University of Chicago Press.

Meyer-Holzapfel, M. 1940. Triebbedingte Ruhezustände als Ziel von Appetenzhandlungen. *Naturwissenschaften* 28:273–80.

Mittelstaedt, H. 1957. Prey capture in Mantids. *Recent Advances in Invertebrate Physiology*. University of Oregon Publ. 51–57.

Pavlov, J. P. 1927. *Conditioned Reflexes*. Trans. G. V. Annep. London.

Prechtl, H. F. R. and Knol, A. R. 1958. Die Fussohlenreflexe beim neugeborenen Kind. *Arch. Psychiat. Z. Ges. Neurol.* 196:542–53.

Richter, C. P. 1954. Behavioral Regulators of Carbohydrate Homeostasis. *Acta Neurovegetativa* 9:247–59.

Riess, P. E. 1954. The Effect of Altered Environment and of Age on the Mother-Young Relationships among Animals. *Ann. N.Y. Acad. Sci.* 57:606–10.

Russell, W. N. 1958–61. Evolutionary Concepts in Behavioural Science: I–III General Systems (Year Book of the Society for General Systems Research), 1958, 3:18–28; 1959, 4:45–73; 1961, 6:51–92.

Schleidt, W. 1961. Reaktionen von Truthühnern auf fliegende Raubvögel und Versuche zur Analyse ihrer AAMs. *Z. Tierpsychol.* 18:534–60.

Schleidt, W. 1964. Wirkungen äusserer Faktoren auf das Verhalten. *Fortschr. Zool.* 16:469–99.

Schleidt, W. and Schleidt, M. 1960. Störung der Mutter-Kind-Beziehung bei Truthühnern durch Gehörverlust. *Behaviour* 16:3–4.

Schutz, F. 1963. Objektfixierung geschlechtlicher Reaktionen bei Anatiden und Hühnern. *Naturwissenschaften* 19:624–25.

Schutz, F. 1964. Die Bedeutung früher sozialer Eindrücke während der "Kinder- und Jugendzeit" bei Enten. *Z. Exptl. Angew. Psychol.* 11:1, 169–78.

Storch, O. 1949. Erbmotorik und Erwerbmotorik. *Anz. Mat. Nat. Kl. Öster. Akad. Wiss. Wien, H.* 1:1–23.

Tinbergen, N. 1952. *Instinktlehre*. Berlin: Paul Parey.

Tinbergen, N. 1955. Some Aspects of Ethology, the Biological Study of Animal Behaviour. *Advan. Sci.* 12:17–27.

Tinbergen, N. 1963. On Aims and Methods of Ethology. *Z. Tierpsychol.* 20:404–33.

Thorndike, E. L. 1911. *Animal Intelligence*. New York.

Thorpe, W. H. 1956. *Learning and Instinct in Animals.* London: Methuen.

Tolman, E. C. 1932. Purposive Behavior in Animals and Men. New York: Appleton-Century.

Wall, W. von de. 1963. Bewegungsstudien an Anatiden. *J. Ornithol.* 104:1–14.

Wall, W. von de and Lorenz, K. 1960. Die Ausdrucksbewegungen der Sichelente (*Anas falcata L.*). *J. Ornithol.* 101:50–60.

Zippelius, H. and Schleidt, W. 1956. Ultraschall-Laute bei jungen Mäusen. *Naturwissenschaften* 43:502.

Author Index

Subject Index

[Italic page numbers indicate definitions.]